WAR IN PRACTICE

SOME TACTICAL AND OTHER LESSONS
OF THE CAMPAIGN IN SOUTH
AFRICA, 1899–1902

BY

MAJOR B. F. S. BADEN-POWELL
SCOTS GUARDS

WITH AN INTRODUCTION BY
MAJOR-GENERAL R. S. S. BADEN-POWELL, C.B.
INSPECTOR-GENERAL OF CAVALRY

WITH DIAGRAMS AND ILLUSTRATIONS

LONDON
ISBISTER & COMPANY, LIMITED
15 & 16, TAVISTOCK STREET, COVENT GARDEN
1903

PRINTED BY
WILLIAM CLOWES AND SONS, LIMITED,
LONDON AND BECCLES.

SPION KOP.

PREFACE.

WAR in theory has often been written about, but practical experience is all we can really rely on in conducting hostilities, and times have changed. After all the experience gained in South Africa (and at what cost!), let us not forget the lessons learned. Time soon impairs our memories, and if we are not reminded of what took place, our impressions will soon be clouded over and details of value forgotten. It is for this reason that I have collected and elaborated my notes, all written actually in the field.

My object is not so much to instruct those who have gained their experience in the war, as to remind them of a few forgotten facts, and to put forward suggestions on certain points which may form wrinkles worth bearing in mind. It may also be useful to offer experience secondhand to those who have not had the opportunity of acquiring it for themselves.

PREFACE.

Since my notes were arranged, some new regulations have appeared; but, though these official works must govern the methods practised, I have thought it best to leave in their original form some remarks which may not be strictly in accordance with those regulations.

Though I had hoped, and still hope, to see some one more competent than myself bring out a complete work on modern tactics and military science, yet it has to be remembered that no one man can be equally experienced in all branches. While a regimental officer is seldom able to grasp the complicated situations in a large engagement, and is often unable to appreciate the difficulties which an officer in supreme command may have to contend with, yet, on the other hand, the general and the staff officer will usually have but little experience of the minor practical details; such, for instance, as those of outpost duty, or the construction of trenches. Having served on and off for some five months in various staff capacities, for several weeks at different times in independent command of a mixed force, and in doing regimental duty for the remainder of nearly three years of continuous active service, I had rather exceptional opportunities of studying the various subjects here gone into, especially as it was my good

PREFACE.

fortune to visit during this time the scenes of nearly all the principal engagements in the Transvaal, Orange River Colony, Natal, and Cape Colony.

It has often been urged that the circumstances of the war in South Africa were so peculiar, and so unlikely to occur again, that the lessons there learnt are hardly worth study. I have endeavoured to bring forward only such points as would seem applicable in all warfare.

Some of the statements here made may not meet with universal acceptance. Opinions are bound to differ when such a subject is treated of. But the drawing-out of criticism is most necessary in properly elucidating the teachings of the war; and the bringing forward of suggestions to thinking men is, after all, the essence of military education, where so few hard-and-fast laws can apply.

My best thanks are due to my brother, who contributes an Introduction, and to Brigadier-General Rimington, both of whom have kindly gone through the proofs and have added valuable remarks.

B. B.-P.

CONTENTS.

CHAPTER		PAGE
I.	Strategy and General Conduct of Operations	15
II.	The Attack	42
III.	Defence	76
IV.	The Selection of Ground and Positions.	108
V.	Fortification	141
VI.	Outposts and Protective Screens	200
VII.	Arms and Armaments	242
	Appendix.—A Typical Position	267
	Index	271

LIST OF ILLUSTRATIONS.

	PAGE
SPION KOP	*Frontispiece*
ARMOURED TRAIN	30
HART'S HILL, SHOWING THE SANGARS	47
CONTINUOUS TRENCHES ON VAAL KRANTZ	90
BOER GUN-EMPLACEMENT ON TUGELA HEIGHTS	102
CHILD'S HILL, FROM THE SOUTH-WEST. SHOWING CONVEX AND CONCAVE SLOPES	110
BOER SAND-BAG FORT: MAFEKING	147
SPRING-GUN ON FENCE	165
GOOD LOOPHOLES (PIET RETIEF)	165
COVERED-IN TRENCH, IN COURSE OF CONSTRUCTION	176
COVERED-IN TRENCH	176
TYPICAL BLOCK-HOUSE	178
TYPICAL BLOCK-HOUSE	178
FLAT-ROOFED BLOCK-HOUSE	180
A LOOK-OUT TOWER	180
AN ARMOURED WAGGON	186

LIST OF ILLUSTRATIONS.

	PAGE
Boer Gun-emplacement	230
Transvaal Artillery before Ladysmith	234
"Long Tom" before Mafeking	250
Colenso: Spot where the Guns were lost	256

LIST OF DIAGRAMS.

	PAGE
Concentration of Fire	83
Posts protected from Each Other's Fire	93
Profiles of Hills	110
Relative Command	119
Position at Base of Slope	120
Diagonal Shelter-pits	152
View from Level of Trench	157
Narrow Loopholes limit Fire	160
Semicircular All-round Trench for Six Men	162
Head-cover	164
Notched Parapet	167
Plan of Loopholes	168
Covered-in Trench	177
A Typical Position	269

INTRODUCTION.

ONE product of the recent war in South Africa has been the Amateur Military Critic, whose writings would lead us to believe that this campaign has introduced vital changes into the art of war such as will govern the conduct of operations in the field for the future. But, as a matter of fact, no new principles of strategy or major tactics have been exploited.

A war in which one side possessed neither cavalry nor infantry nor a numerous artillery, but was practically composed entirely of mounted irregulars, without definite strategic plan or systematic discipline and direction, could hardly be expected to effect any great alteration in the strategy of the future.

In tactics, however, the experiences of the campaign, even if they taught no new principles, nevertheless had their use in confirming and putting the seal of war-trial on the ideas of men

INTRODUCTION.

who had studied the theories of war, and they were valuable in brushing away errors and anachronisms, and also in giving us numerous minor hints for application to our future work and training.

Such points being small in themselves, are apt to get lost sight of in settling down again to peace methods, unless made note of at the time.

In the following pages, written by my brother while still on service in South Africa, will be found the application of experiences of the campaign to the practice of modern tactics in the field. They may, therefore, be of interest or profit to peruse. For, though it may be said that success in war is the outcome of the three C's—courage, commonsense, and cunning—yet experience has also a good deal to say in the matter; and no officer, even the *youngest* of us, has so great experience that he may not with advantage supplement it by the experiences of others. In fact, therein lies a great preliminary step towards success, viz. in the STUDY as well as in the practice of one's profession.

R. S. S. BADEN-POWELL.

April, 1903.

WAR IN PRACTICE.

CHAPTER I.

STRATEGY AND GENERAL CONDUCT OF OPERATIONS.

WAR is a very old-established institution in the life of man. The means by which it is waged, the principles of strategy and tactics, are founded on experiences of the past. Hence the subject must be a conservative one, and cannot undergo sudden changes. Yet if old ideas be persistently clung to when new circumstances prevail, failure and disaster may result. The tactics of to-day are but the outcome, slowly developed, from the days of bows and arrows, and there are many institutions which we still cling to which would certainly not have been introduced had a system been founded at one time and not evolved from past ages. *Evolution of strategy and tactics.*

Strategy is perhaps less affected by improvements in armaments than are the more minor operations of war. Neither effective range of weapons, rapidity of fire, increased accuracy of fire, invisible discharge, or, in fact, any other such innovations, affect the question. It chiefly requires the simple application of common sense to the circumstances of the moment ; which circumstances are so variable that no very strict rules can apply. The manœuvring of troops in the presence of the enemy, on the other hand, is more amenable to preconceived laws, and has to be modified by considerations of his armament and newly introduced conditions on the battlefield. Oft-repeated experiences under more or less identical conditions, will, from time to time, necessitate some alterations in these laws.

And yet the South African War presents several points which open up new ideas on the larger operations ; and by the consideration of past experiences we may open our minds and be better able to grasp difficult problems set before us in the future.

Invaders and defenders. In considering the broad principles of any war, the opponents may usually be classified as invaders and defenders. The one force will be advancing through the country, endeavouring to gain the capital town or seat of government, or to follow up and defeat the enemy's army. The other force will be more or less passively trying to defend its positions on the frontier or on interior lines—that is to say, not attempting any general forward moves.

The assailant has the great advantage of being

able to choose his point of attack. By keeping plans secret, and organizing feints and diversions, he is able to concentrate an overwhelming force against any spot which he considers to be the enemy's weakest, either to break through the line of defence, or to inflict a defeat, and so weaken the hostile forces.

When Sir Redvers Buller arrived in South Africa with his Army Corps, the troops of the respective nations were practically at a standstill, facing one another in several chosen positions. Buller then was in the position of an invader, able to choose his point of assault. Whether, therefore, his best course was to divide up his forces and push them up to reinforce the several places already held, with the hope of pushing each forward, or to concentrate his whole strength in some direction where he found the enemy to be weak, were the main points for him to decide. He chose the former. But in this instance, as is often the case in strategy, there were other considerations. The hordes of Boers, flowing into Natal like a rising tide, had swept around Sir George White at Ladysmith, hemmed him in, and were now spreading south, threatening to invade lower Natal. For political reasons it was desirable to stay this invasion. Whether a determined advance into the enemy's territory from another direction would have had the desired effect is a moot question. Still, here is an instance of the invader not attacking in a weak point, and the result was, that beyond staying the invasion nothing was accomplished until events elsewhere affected the position.

Later on, when Lord Roberts arrived, the general situation was again the same. Then, however, a secret advance was made in an unexpected direction into the enemy's territories, which necessitated the immediate withdrawal of the defender's forces from all his strong points.

Invader's advance. The assailant's advance may be from one point, or from several points. In the latter case there will usually be a better chance of success, since it will be easier to find out the weak spots of the defence, and if one of the columns succeeds in breaking through the general line of defence, it can take other positions in flank and rear at the same time that they are being attacked by other columns in front.

Thus at the end of 1899 there was presumably an intention of advancing our forces by several routes, through Natal, Stormberg, Colesberg, and Kimberley. As it turned out, each happened to be checked; but had one of them succeeded in breaking through—for instance, had Lord Methuen been able to advance into the Free State, he could have threatened the rear and communication of the enemy to the south, and so assisted Generals French and Gatacre to advance.

On the other hand, this invasion by several columns necessitates a dividing up of the invader's forces; and if the several advances are made on points far separated, the columns are unable to

GENERAL CONDUCT OF OPERATIONS. 19

give one another mutual support. This would undoubtedly be to violate a well-recognized principle in war; that is to say, a force, divided up into portions, *unable to afford one another mutual assistance* (and not being utilized merely as tentative skirmishers to seek a weak spot in the defence) is, as a whole, unable to bring that overwhelming force to bear which is necessary to crush the enemy at a desired point.

Sir R. Buller, by dividing up his forces, sending one quarter of his infantry towards Kimberley, half his cavalry to Colesberg, and retaining only about one half of his Army Corps for use in Natal, found himself not strong enough to advance anywhere; whereas, had he been able to concentrate the greater part of his strength at one point, it is more than probable that, at all events, a local victory would have resulted. This is given merely as an instance. Under the actual circumstances political reasons may have necessitated the dispositions adopted.

Improvements in modern firearms have added so greatly to the power of defence that the invaders, to be successful, must be in greatly superior force, whether this be in numbers, armaments, or otherwise, so that with several invading columns unable to give mutual support, each must be very strong indeed to succeed. But each of these columns cannot be as strong as the whole combined, and are therefore more liable to repulse. With the long range of modern firearms, however, separate forces can support one another better than formerly. Thirty years ago columns only separated by one mile could render one another but little real assistance, whereas now, with heavy field-guns carrying over six miles, and maxims and rifles firing bullets accurately to nearly two, at least three or four times that distance may intervene, and yet they can co-operate.

On the other hand, if the advance be by one line only, and if the defenders have managed to ascertain the direction of it, a strong defence may have to be broken; and as frontal attacks have been found to be undesirable under modern conditions, wide turning movements and strategical flank advances will often have to be undertaken, which is practically equivalent to several points of advance.

The value of pushing forward on the flanks seems scarcely to have been fully recognized by our generals in the early stages of the war, though clearly proved by Lord Roberts's successful advance later on.

Flanking advance. It may be contended that such wide outflanking movements would necessitate the employment of a comparatively very large force (herein presumably the reason of their not being carried out in these cases), for if but a weak force, or none at all, be left facing the enemy, a counter-advance might be made, cutting the communications of the invaders.

But we have now learnt, what was not at that

GENERAL CONDUCT OF OPERATIONS.

time fully realized, the feasibility of a small force occupying a strong position to repel any such counter advance. If, after the battles at Belmont and Grasspan, a few hundred men had been left strongly entrenched about those places, while a turning movement to the eastward was in progress by the rest of Lord Methuen's force, the Boers could hardly have advanced with any hope of successfully cutting off our forces from their base; while the movement would probably have forced the enemy to retire from his position at Modder River. So in Natal, if a strong position had been held on the railway, a wide and rapid turning movement could not have received very serious opposition without allowing the Ladysmith garrison a great opportunity of breaking through south. For if the movement were sufficiently *wide* and *rapid*, the Boers could not have taken up an entrenched position and be ready for our attack, as they were.

Looking at the subject now from the defender's point of view. To avoid such "points of defence" being thus turned, long "lines of defence" may be taken up. The Boers, having failed to stay our advance on Kimberley by the positions at Belmont, Grasspan, and Modder, occupied and entrenched at Magersfontein a line of such a length (over ten miles) as to be practically impossible for the force then advancing to turn, at all events in a tactical sense. Such long "lines of defence," stretching for many miles of front, will probably be largely

^{Lines of defence.}

utilized in future warfare. Improved firearms (and the tendency will be to emphasize the newest characteristics of improvement) enable long lines to be held with comparatively small forces, and really be sufficiently long to be unturnable.

These lines would, of course usually be made on natural or existing features. Thus a large river at once forms a good basis for defence, and the Boers might have made a most formidable opposition had they utilized such. They might have prepared a long length of the Vaal to oppose our advance on Pretoria, and, if well entrenched, it would have been practically impossible to turn them out. A big river is particularly suitable for this purpose, as fords and bridges can be strongly held; the unfordable remainder only lightly held by very extended forces. And rivers can be readily converted into formidably impassable barriers.

A good line can be taken up in front of a railway, the latter not only facilitating the distribution of supplies, but could also transport reinforcements to any threatened point.

Such strategical lines of defence need not form the frontier of the country; indeed, it will often be preferable to withdraw back some way, especially if a sudden outbreak of war finds the country unprepared. In Natal we retired back to Ladysmith and behind the Tugela, instead of holding the frontier. This necessitated a long advance for the enemy, giving us time to organize our defence; and, had matters been in a better state of

GENERAL CONDUCT OF OPERATIONS. 23

preparation, we could doubtless have very greatly prolonged the time of that advance.

Had the strategy of the Boers, after the occupation of Bloemfontein, been to occupy the line of the Vaal, they could have had ample time to render it a very strong position, and by harassing our northward march as much as possible with small bodies of mounted troops, could have delayed the advance until the defences were complete.

It is evidently desirable, under most circumstances, to force an enemy to traverse a considerable extent of ground before coming to decisive fighting. Smokeless powder and long range firearms have rendered reconnaissance so extremely difficult, that no advance can be made rapidly if there are even a few defenders continually occupying successive positions, and falling back when pressed. If the country be much enclosed or broken, reconnaissance, and consequently rapid advance, will be still more difficult; and the obstruction and destruction of roads, railways, and bridges will still further retard progress. *Force enemy to traverse distance.*

In his long advance the invader may thus be harassed and delayed; his forces weakened by men left to guard communications; his men, horses and transport animals more or less exhausted, and meanwhile plenty of time is available for the defenders for preparing defences and making dispositions for the fight. It also affords them opportunities of ascertaining the strength and dispositions of his adversary.

Lord Roberts, by making his main advance from Modder River, far away from the main Boer army around Ladysmith, forced the latter, or such portion of it as could be spared, to traverse right across the Free State and, in an attenuated form, to come to new ground.

Invincibility of small garrisons. One of the great lessons of this war is the proved ability of detached garrisons, in rapidly improvised fortifications, to hold out for very considerable periods against not only the onslaught of overwhelming bodies of the enemy (for this may be ascribed in part to the want of dash and cohesion of this particular enemy), but against the bombardment of many most powerful guns. The sieges of Mafeking, Kimberley, Ladysmith, and other smaller events at Wepener, Elands River, Ladybrand, Lichtenberg, etc., prove this most thoroughly.

Had it been realized at the beginning of the war how powerful such defence is, and how small a garrison may keep a much larger force employed, we would doubtless have held on to many places which we abandoned. If in Natal positions had been taken up about Dundee, Glencoe, and elsewhere, the force which we held at those places, distributed in, say, four different centres of defence, would probably have stayed the invasion of Natal, and, absorbing a large portion of the enemy's force in besieging them, have left Sir G. White's force practically free to operate against the remainder.

The uselessness of investing such places is also to be noted. Had the Boers left but small

GENERAL CONDUCT OF OPERATIONS. 25

containing forces around Mafeking and Kimberley, they could have added several thousand men to their forces in the south.

Such detached garrisons would, too, form valuable stepping-stones to facilitate an advance whenever that could be undertaken. It would always be far easier to relieve a garrison holding a strong position, than to attack and drive an enemy from that position.

When small bodies of troops are placed in isolated positions without due preparation, endeavours should always be made to have some supporting troops near. A small force can always give a good account of itself; but if from not being able to locate itself in a good position, or if arrangements for supplies of food or of ammunition have not been made, or from other causes, the limit of time it can hold out may be greatly reduced. If it be within a few miles of another post, a message can generally be got through to state the position of affairs; or the firing of an attack may be heard, and in this way help may be sent.

Support to isolated forces.

The five companies of Northumberland Fusiliers and Irish Rifles, retiring from De Wetsdorp, were surrounded by a superior force, and, after holding out all day in a position where they were attacked, were forced to surrender, they being deficient of supplies, and there being no troops near to succour them.

So at Uitval Nek; the Greys and the Lincolns, though they had warning of impending attack, and though within 20 miles of the forces occupying Pretoria, had to surrender after a fierce fight lasting all day. Many other instances of the same kind occurred later in the war. Had quite a small force been at hand, able to pass on the alarm, or even to demonstrate at a relief, it is quite probable that the detachments would have been saved.

Advance independent of base. An invader, finding it necessary to make an advance through a considerable tract of enemy's country, may deem it best to sever connection with his base, and thus not be fixed to one route or obliged to guard his line of communications. This gives him great independence, and leaves his forces intact, but, on the other hand, cannot be undertaken unless very confident of success and a reasonable certainty of renewing that connection before supplies become exhausted. Lord Roberts's great advance from Modder River to Bloemfontein was a wonderful instance of this. He took an army of 40,000 men 100 miles into the enemy's country, driving a large mounted force before him all the way. It took six days to reach Bloemfontein, where he counted on the enemy evacuating the railway line southwards so as to enable him to revictual by that route.

Lord Roberts's own words, used in his despatch written before the actual advance, clearly describe the advantages of being free of a line of communication :—

"The two main roads leading from Cape Colony

to the Orange Free State were held in force by the Boers at the points where those roads crossed the Orange River, and it seemed certain that the bridges over that river would be destroyed if the enemy could be forced to retire to the northern bank. Moreover, I could not overlook the fact that, even if either of these routes could be utilized, the movement of an army, solely by means of a line of railway, is most tedious, if not practically impossible. The advantage is all on the side of the enemy, who can destroy the line, and occupy defensible positions when and where they please. In a hilly, enclosed country, or where any large river has to be crossed, they can block the line altogether, as was proved in the case of Lieut.-General Lord Methuen on the Modder River, of Lieut.-General French on the Orange River, and of General Sir Redvers Buller on the Tugela.

"A railway is of the greatest assistance; it is indeed essential to an army for the conveyance of stores and supplies from the base, and it is a most valuable adjunct if it runs in the direction of the objective; but even then a certain proportion of the troops must be equipped with wheel or pack transport to enable supplies to be collected, and to render the force sufficiently mobile to deal with many tactical difficulties which have to be surmounted, owing to the greatly increased range and power of modern projectiles."

While speaking of the dependence of a moving column on a railway for supplies, a very practical instance must be noted. When the advance of the relief force for Mafeking was decided on, the question of transport was a serious consideration. Transport animals were very scarce, and the very best would be necessary for such a rapid movement. As an alternative idea, it was suggested that the railway might be used. It was certain that many of the bridges had been destroyed and the line elsewhere damaged, but it was thought that by using light trolleys drawn by mules, and carrying a certain amount of material for repairs, these obstructions could be got over. In the end, however, the requisite number of mules were procured for road transport, and the railway not used. This was very fortunate, as it was afterwards found that the rails in one place were torn up for about seven miles on end!

During the later phases of the war small mobile columns carrying their own supplies were the order of the day, and were generally most successful.

The success of such columns may, however, be easily marred by adverse circumstances. Thus, had wet weather come on during Lord Roberts' march (and torrents of rain had fallen but a few days before), it would have been extremely doubtful if the ox-waggons could have accomplished the march, and a long delay *en route* might have proved very serious.

At Piet Retief several of General French's mobile columns found themselves unable to progress or obtain supplies, through the rivers and roads be-

GENERAL CONDUCT OF OPERATIONS. 29

coming impassable from heavy rains, and the troops had to go on very short rations in consequence.

One of the most important points to consider in the advance of an army into the enemy's country is the line of communication and its protection. **Lines of communication.**

As, however, the power of defence has increased, it should now be easier than formerly to protect the communications. There was certainly plenty of experience in guarding long lines of communications in South Africa. Not only had hundreds of miles of railway line—eventually every line in the country—to be guarded, but also many main roads along which supplies had to be conveyed to more distant garrisons.

Railways, without doubt, form by far the most efficient means of supplying an army, but they are, unfortunately, very vulnerable. On hundreds of occasions the rails were blown up by small mobile parties of the enemy, causing much delay; and the very numerous bridges had to be carefully guarded.

Ordinary roads are much more difficult to render impassable to waggons; but then the latter are so slow and cover so much ground that they form a tedious method of supply. Traction engines modify these objections, and perhaps in the future improved motors will greatly facilitate the question.

In a general advance along a railway it may be desirable to move a number of trains close together, bringing a large store of supplies, so that if the line be broken behind, there will be plenty to go on with; thus there would be no difficulty of guarding the lines of communication. In advancing over a dangerous bit of railway I have known six trains to run together only a few hundred yards apart.

If operations are taking place away from the coast, railways are of the greatest importance, and it is of course preferable in case of one being cut off, or temporarily destroyed by the enemy, or accidental causes, if several such run towards the field of action.

Concentration and extension. It used to be said that a force must *concentrate* to fight, but *extend* to obtain supplies. Recent experience rather tends to the opposite. Thus the tendency in modern fighting, from the individual man to the large force, is to extend. The separation of larger units enables the commander to "work" the country better; to reconnoitre on the very large scale now necessary; to outflank positions taken up by the enemy; to prevent being outflanked; and yet the long extended lines are, owing to the great range of modern firearms, able to offer considerable opposition to attack. The closing in of various units may of course be necessary for a definite assault or combined action. .

But, at all events in a country like South Africa, where we were almost entirely dependent for supplies on the railway or from certain depôts, it became necessary, or at least desirable, for all the forces occasionally to come near to that base for their food—hence to concentrate for supply.

The advance through an enemy's country in a number of comparatively small, independent, but

GENERAL CONDUCT OF OPERATIONS. 31

mutually supporting columns, has been proved to be the best system for clearing up the small hostile bands remaining, or even for feeling the way in very extensive territories; similar in principle to the scattered individuals in tactical skirmishing.

Though in theory the working of such extended forces may appear as easy as directing a line of skirmishers, in practice great difficulties were shown to exist. In the case of South Africa, lack of good maps, paucity of roads, difficulties of communications and of supply, rendered the control more arduous than ever. The result was frequent want of system and loss of cohesion between the parts; so that small columns were wandering aimlessly about, unsupported, ignorant of the position of neighbouring forces, and wanting definite orders. The attempt to control movements from headquarters hundreds of miles away often proved fatal to some good scheme organised on the spot. Thus a force of the enemy was sometimes located by a column, and the information communicated to neighbouring columns. When perhaps three or four columns were converging on the hostile body, sudden orders would come from headquarters to move off in some other direction, and so opportunities were frequently lost.

The breaking up of a large force into a number of small independent columns may often be desirable for other purposes. *Small and large forces.*

A small force has certain great advantages over a large one.

1. It is easier to feed on the country, and to supply.
2. It is less dependent on good roads, its comparatively small transport not wearing them out, or tailing out so long.
3. Easier to protect with advance and rear guards; also, as a rule, with outposts.
4. More mobile. A large force is always liable to delays.
5. It numbers a larger proportion of fighting men, since a large force necessitates a large staff of extra officers and their attendants.
6. Can be easier controlled and directed by one head.

Strategical attack and tactical defence. As defence has been proved to be so much more powerful than formerly, it would seem desirable to advocate, as the general policy of a campaign, always to act on the defensive. But then, it may be retorted, how can progress result? Strategical advances can, however, be made beyond the reach of guns and rifles, and when actual combat becomes imminent, a tactical position of defence can be taken up. This, broadly speaking, was done by the Boers in Natal.

It used always to be laid down that passive defence was useless in bringing about any decisive result, but, if it be the case that defence has increased in power, it must now be of more value. If combined with strategical advance, it may be proved to be even of considerable

GENERAL CONDUCT OF OPERATIONS.

importance. The great object would be to so manœuvre as to cut off the enemy from his base, and to force him to attack if he attempts to regain possession of his communications. This would practically, as a rule, mean surrounding him. In this way Cronje was headed off at Paardeberg, surrounded, and forced either to attack and break his way out, or to capitulate. With reference to this instance, it may be added that we also received a very severe lesson as to the uselessness of attacking. The assault resulted in the most disastrous battle of the whole war, and did us no good whatever. All that is necessary is to get round the force, and take up passively defensive positions, such as to prevent its retreat, and to prevent supplies coming to it.

This may prove of special importance in European warfare—where, if a few well-chosen positions on all the roads leading to where the enemy is be strongly occupied, it may be sufficient to eventually force him to surrender.

In this manner a small force may even surround a larger one, and defeat it.

It used formerly to be considered that the ruling feature of the battlefield, the ultimate aim of all manœuvring, was to bring the opposing forces into actual contact. It has often been said that by fire alone no decisive results can be achieved. But this is another principle of war which has undoubtedly undergone a change. The two operations having the most decisive results in the whole

Shock tactics not essential.

South African War, the surrenders of Cronje and of Prinsloo, were brought about practically without any hand-to-hand fighting, and, by skilfully manœuvring to cut off an enemy from his base, and by bringing a continuous heavy fire upon him, he will be bound to come to subjection. This principle applies equally to smaller events, such as the capture of Col. Carleton's force at Nicholson's Nek, and several other affairs when small bodies of our troops were surrounded by the Boers, and to many captures of commandoes in the later part of the war.

Retirements objectionable. It must be recognized that retirements from a place once occupied are usually a mistake. The retreat from Dundee, where the whole force had to precipitately abandon the place on the approach of a greatly superior force of the enemy, leaving stores, kits, and even wounded, is a regrettable instance.

Such retirements before approaching hostile forces are demoralizing to the troops, encouraging to the enemy, and render the force exposed to attack while on the move. It could easily be avoided by proper disposition of troops.

In a well-thought-out campaign, a position considered to be untenable, either because of weakness, unsuitability of ground, or want of supplies, if attacked by a force such as the enemy might be likely to bring against it, should be relinquished in plenty of time, or reinforced and strengthened.

Several times during the campaign towns had to

GENERAL CONDUCT OF OPERATIONS. 35

be suddenly evacuated after more or less prolonged occupation, and though the retreat may not have been as precipitate as in the instance just mentioned, still the objections were all manifested and the troops much disheartened to see, as it were, all their work of conquest undone.

Rustenberg was an instance where the town was occupied, very strongly fortified, and a great amount of work done to make the place healthy and comfortable, when a sudden order came to evacuate it at once.

Retirements give the enemy a good opportunity of pursuit, during which much damage may be inflicted. The method of pursuing a more or less defeated army is a subject to be carefully considered. The most natural and usual way is to follow on the heels of the fleeing foe, but this method is open to several objections. The pursued are then able, by skilful dispositions of their rear guard, to delay and even inflict damage on the pursuers. Opportunities may often occur of practically stopping the pursuing force, as, for instance, by crossing a bridge and then destroying it. *Pursuit.*

A preferable method would usually be to follow by routes on the flank, and, if the enemy's destination be evident, an even shorter route may be utilized. For instance, when Cronje retired from Magersfontein, it was evident that he would make in the direction of Bloemfontein. Having crossed to the north of the Modder River, it was clear that he would have to again cross to the south bank, and, therefore, instead of following him across the intricate "drifts," the bulk of our forces moved along the south side of the river, and were able to head him off at Paardeberg.

Rapidity. Rapidity of advance is a most important characteristic in many operations. It may be said to depend on mobility and on preparation. Several good instances might be quoted in which Lord Roberts delayed for a long time in one place, making preparations, and then, when all was ready, making a rapid, telling, and always successful move. Had he steadily pressed on at a slow rate there can be no doubt but that the difficulties of supply, etc., would have been increased, and the enemy better prepared to oppose the advance.

So, also, the Boers on several occasions rather surprised us by suddenly turning up in large numbers at an unexpected point, having made a long and rapid march.

Some of our movements in Natal would undoubtedly have proved more successful had it been possible to carry them out with greater rapidity.

Yet, undue haste, and advancing without full preparation, is not desirable. When, towards the end of the war, it became necessary to clear the country of supplies, the columns were sometimes unduly hurried, and were unable to effect their work with proper thoroughness, with the result that it all had to be gone over again.

To go unduly prepared is to court disaster.

Surprise. One of the chief objects, as a rule, in a rapid movement, is that the enemy may not expect it, that he would not calculate on the force being at that particular place so soon, and so, in fact, could be surprised.

Surprises, in war, have the greatest effect, whether in the extensive movements of armies, or in the action of two or three men. Whether it be the sudden appearance of an unexpected body of the enemy, his absence where expected, his presence in greater strength than anticipated, or an unexpected attack, or an ambush to stop an advance—all have a most demoralizing effect.

Almost every one of our reverses in South Africa can be ascribed to our being taken unawares; and many of our successes, too, are due to our surprising the enemy.

An attack on a force expecting it can seldom be successful, while a secret and unexpected dash on a camp or post will have a good chance of accomplishing its object.

It was most noticeable how often some operation which was to be conducted in the greatest secrecy was anticipated by the enemy, and, which was the reason, known about in our own camp, long before the orders for its execution were given. It is said that the night attack on Stormberg was discussed in camp several days before its abortive accomplishment. So, a certain night march to Roossenekal, to surprise the Boers who had gone back there, was known and talked about, by officers and men alike, the day before. I happened to speak of it to the Chief Staff Officer, who was intensely surprised to hear of its being known, as the idea was supposed to be a secret between the three or four senior officers.

Allotment of troops. The allotment of suitable troops to the various localities is a subject deserving of more attention. The bottling-up of a brigade of cavalry in besieged Ladysmith has often been adversely criticized, and there are many other instances in the campaign similar, though often, perhaps, unavoidable. Thus, the force composed entirely of irregular mounted corps under General Baden-Powell, operating from Mafeking towards Pretoria, found themselves hemmed in at Rustenberg, and not being strong enough to force the enemy away, and hold the several positions necessary at the same time, were forced to remain practically inactive at the latter place for some months, while mounted troops were urgently being called for, for duties elsewhere.

So, on the other hand, there are numerous instances of places being garrisoned entirely with infantry, where a few cavalry were much needed for active work in the neighbourhood.

It should always be a matter for consideration where the capture of a town is contemplated, as to how it is to be garrisoned.

Night operations. Night marches may be undertaken for various reasons. (1) For secrecy, such as the retreat of the force from Dundee. (2) For rapidity. The Mafeking relief force, making its forced march, travelled as much by night as by day. (3) To

GENERAL CONDUCT OF OPERATIONS. 39

approach a position unseen, and not to come under long-range fire. The first may be a necessity, and will often be desirable to effect a surprise attack. The second is often most advantageous, if circumstances are favourable, such as a moonlight night. But to march along a bad road in pitch darkness involves delays and checks which render progress very slow and very tiring, so that it would, in most cases, be better to wait till daylight, when probably three times the rate of progress can be maintained. On one occasion, marching from Pienaars River, we started on a fairly fine night, but clouds gathered over and a violent storm came on. The road was bad, with bushes at the sides, into which we constantly fell, checks were frequent, and the march was very slow and most exhausting.

Large bodies of troops, especially if accompanied with guns and transport, can seldom make a successful march on a dark night. They certainly cannot move secretly. The rumble of the wheels can be heard for miles, and among many thousands of men there is sure to be an occasional match struck or a rifle accidentally discharged. Such occurred on our night march to Magersfontein. Checks were frequent, and several units lost their way.

Small picked bodies may often accomplish much good at night, especially if they have not far to go. The attack on the Boer trenches at Mafeking, and the blowing up of the Long Tom at Ladysmith, are instances.

Such small parties may often be profitably employed in disturbing the rest of the enemy before an attack is delivered. After alarms all night, the defenders are not in the fittest state for repelling attack next day.

Guns, aimed by day, and fired off during the night, often caused considerable alarm, if nothing more.

One of the great objects of a night attack will be to create a panic. Men suddenly aroused from sleep are very liable to be panic-stricken, and may then do foolish things.

Fog.

Fogs by day have characteristics differing from mere darkness. (1) Those familiar with the ground and with its minor landmarks—footpaths, bushes, and so on—will be able to find their way about, while a stranger may be utterly at a loss as to his direction. (2) Rifles may be sighted and used at close quarters, and even beyond the range of vision by pointing in any desired direction, quite as well as in ordinary daylight. (3) Good cover can be selected and even constructed. (4) Friends can be recognized from foes. These characteristics all favour defence. An assailant, not knowing the ground, will have great difficulty in finding his way, and in realizing what is going on around him, while the defenders have but to lie still and shoot in what they know to be the right direction.

Summary.

The points, then, in which strategy is affected by recent changes may result in :—

1. Tactical defence having increased in power in

a greater ratio than attack, strategical plans must rather aim to so move the forces as to bring them into good defensive positions, rather than to move them so as to be able to attack advantageously.

2. Shock action is becoming of less importance, and the object of manœuvring in the field will be to force the enemy into subjection without necessarily coming to close quarters.

3. Owing to the increased range of firearms, especially heavy field artillery, the area of tactical operations is greatly enlarged, and very long lines of defence can now be taken up such as were formerly impracticable.

4. Modern innovations tend to favour small forces, so that, as a rule, a large force is better disseminated into a number of small ones, mutually supporting one another.

5. Reconnaissance now being so difficult, advances in an enemy's country must be slow, and may be easily much delayed.

CHAPTER II.

THE ATTACK.

Attack and Defence. RECENT improvements in firearms, both in rifles and in artillery, have been proved to influence greatly the relative value of attack and defence. Nearly all tend to strengthen the power of the defence.

The reasons may be summarized as follows—

1. Fire being opened at a much longer range than formerly implies that the troops would, for a much longer time, be exposed to its influence, and suffer more accordingly.

It is evident that if 1000 yards of open ground has to be crossed by the assailants, it will take them fully ten minutes to cross. It does not matter how often they halt or for how long, but that ten minutes *must* be spent by them advancing, fully exposed, and unable during that time to fire back. In fact, it means that the defenders can, for periods amounting in all to ten minutes, fire at the fully exposed foe, and in this time each man could fire over a hundred rounds. Even during the time when we may suppose the attackers are lying down in the open, or behind such cover as they may happen to come upon, they will still form a more vulnerable target than the carefully posted, perhaps intrenched, defenders.

2. Musketry fire being so much more rapid, more loss will be inflicted in a given time. And this particularly applies to magazine fire, and where a good target presents itself, into which a mass of fire can be poured.

3. Quick-firing field-guns, too, will have a like effect.

4. The attacking troops cannot clearly locate the position of their enemy, who will, therefore, not suffer as much as when puffs of smoke clearly showed the exact position of each defender, while the attackers advancing across the open will be as conspicuous as ever.

5. The flat trajectory of the modern rifle enables the defenders to sweep the ground more effectually, as the sighting is not so important, and a bullet which misses a man in front may strike one following in rear.

6. Machine-guns can open such a terribly rapid and accurate fire on advancing troops as to make certain zones practically impassable.

In addition to these effects of modern firearms, other conditions, more or less novel, benefit the defence. Cover, now of such importance, can be selected at leisure by the defensive side, but is often entirely unobtainable by the attackers. The latter have to waste a considerable time in advancing,

THE ATTACK.

during which the defenders can keep on firing. The attacking force must be considerably extended, while the defenders can, if necessary, lie shoulder to shoulder behind their cover. The supply of ammunition, now so difficult a question in the attack, is easily arranged for those remaining in or about one position. The defenders may know the ground and ascertain ranges beforehand. The wide extensions now necessary favour those who fain would hesitate to advance, and their fire, such as it is, may thus be lost.

Yet another minor advantage of defence is that, as a rule, supplies of food and water can more easily be arranged for. Actions now last so much longer than formerly that this becomes an important question. Though a man may easily be able to exist, and fight too, without food for twenty-four hours (we were thirty-six hours at Magersfontein with nothing more than what we had in our water-bottles and haversacks), yet, even after six or eight hours, when hunger begins to tell, men's spirits become depressed and energy slackens; and these are two most potent factors in successful fighting.

All this, then, renders the attack generally infinitely less likely to succeed than formerly, and if it be conducted across the open in broad daylight and against well-posted troops, practically impossible of success.

Modder River, Stormberg, Colenso, Magersfontein, Spion Kop, the assault on Paardeberg, and many other engagements, could be quoted as proving this.

On the other hand, there are certain circumstances under which a deliberate attack may still be conducted with success. These are—

When attack is likely to succeed.

1. If the defenders can be *surprised*, and close quarters attained before they are aware of the presence of the assailants.

2. Against a *very inferior enemy*, whether in numbers, in quality, or in armament, which includes having but a limited supply of ammunition. (This may be local, such as pressing home an attack on one portion of a long position, clearly ascertained to be weakly held.)

3. When the defenders are in a naturally *bad position*. That is, if there be cover to advance under, enabling the assailants to approach close to the position, or to rest in safety between forward rushes. Or if the ground be otherwise unfavourable to defence. Also if the defenders are in such a position as to be clearly discernible, and not well protected.

4. If darkness or the nature of the country allows of an *unseen advance* to close quarters (even though expected).

Attacking a force not intrenched or behind very good natural cover should also prove successful *if* a large amount of fire—artillery, rifle, and especially machine gun—can be brought to bear on the position. A steady shower of bullets and frequent bursting of shells make it practically impossible for the defenders to keep up that deadly fire which prevents the attack from succeeding. But this is practically the same as their being in a *bad position.* Against troops well intrenched, however, it is a different affair, and we have learnt to our cost that artillery preparation for infantry attack against such is of but little avail.

If none of the above conditions prevail no attack should be attempted, except perhaps under most desperate circumstances. Whereas, if it be ascertained that several of them are existent, as would frequently be the case, an opportunity occurs which should not be missed.

Instances of successful attacks.
A good instance of a successful attack, when most of the above conditions prevailed, is Belmont. Other successful attacks during the war can all be classified under these headings. Talana Hill was a bad position, being steep and rocky, with dongas and trees at the bottom, and stone walls up its sides, which afforded much cover. The charge of the Naval Brigade at Grasspan only succeeded with considerable loss, and against inferior troops, and backed up by an overwhelming artillery fire. At Zillikat's Nek, too, our troops were very badly posted. At Reddersburg and at Nicholson's Nek our troops ran short of ammunition. At Driefontein our loss was great, and though our opponents were better men than those usually encountered during the war, being Johannesberg police, they were probably not very numerous at that spot.

THE ATTACK.

When an attack has been determined upon, it **Objective** ought to be clearly decided what the objective is **in attack.** to be; but this is not always easy to settle upon. In South Africa it was very rarely that, at the commencement of an attack, we knew within 400 or 500 yards the exact position of the enemy. Formerly we used to be taught to advance up to "the position" and storm it. But we have now discovered how very often there is no one special position. The enemy may be aligned across a given piece of country. But as the attacking force approaches the defenders may fall back, and take up successive positions in rear; or, as we approach the enemy, it may be found that he is not drawn up in one line, but occupies a series of small positions at varying distances, and often not in a line parallel to that of the attack (*vide* map, p. 269). The great object will be to gain a succession of advantageous positions whence a telling fire can be brought on the defenders. It is, however, a matter of the greatest difficulty for a leader under modern conditions of the battlefield to grasp the position beforehand, and give clear orders to the various units. Preliminary reconnaissances seldom procure the desired information. So little is to be seen from a distance, and the extent of ground occupied so impossible to determine, that much must be left to chance. And this is one of the great drawbacks to the attack.

In carrying out a general attack on a position, it is of course desirable to concentrate as much force as possible on what is considered the weakest point of the defence. And it must always be borne in mind that concentration of force practically means concentration of fire.

But at the same time it is very necessary that feints and demonstrations be made on other points, so as to keep the defenders in ignorance of one's real intentions; a military maxim of the greatest importance, but often apparently forgotten by commanders.

The usual plan now carried out is to advance on a very broad front, and, while engaging the enemy all along, to throw the flanks forward so as to endeavour to hem in the force. It may be said that whereas formerly an attacking force made a frontal advance and threatened the flanks, now it demonstrates in front and concentrates its attack round the flanks.

In determining the particular objective for attack, the flanks are always considered the most desirable. It may not now, with greatly extended lines and numerous detached forces, be so easy to enfilade and "roll up" a line of defence as in olden days; but if the flanks can be threatened and kept engaged, the attack may be continued round to the enemy's rear, where he may not be so well prepared for defence, or even if he is, his line of retreat can be cut off and the whole force surrounded.

By surrounding the enemy, the attackers gain a better opportunity of finding weak spots in the position, since such are more likely to be found in the flanks and rear than in the front.

THE ATTACK.

In a natural position, or one not well intrenched, the defenders may suffer from a fire coming from their flank or rear which may nullify the advantages of their cover.

Such surrounding of a force, even without seriously attacking it, would usually be a most desirable achievement, as the aggressors may then take up positions of *defence* around, and forcing the defenders to try to break out, may thus attain that strategical advance and tactical defence which is so advantageous.

It must be remembered that the object is to defeat the enemy and not merely to drive him from his position.

It seems hardly necessary to point out the immense advantage of a familiarity of the ground to men making an attack. Yet on several occasions a position well known to some of our forces had to be attacked, but these troops were not employed in the attack. Thus, in the unfortunate affair at Stormberg, the Berkshires had previously known well the position occupied by the Boers, and had actually made works there. When the time came for us to attack the place, that regiment was unfortunately left behind. The same could be told of the reoccupation of Zillikat's Nek, the attempt to relieve the Elands River post, and many other affairs. *Knowledge of the ground.*

Daybreak has always been considered as the most favourable hour to carry out an attack, and successes on many occasions bear this out. On *Hour for attack.*

THE ATTACK.

the other hand, night marches on a large scale (a more or less necessary preliminary) are only desirable under certain conditions (referred to on p. 38).

Small parties sent in advance may, however, do great work in preparing the way. An attack may be conducted late in the afternoon, so that the decisive zone be reached just at dusk, and the final charge made after dark. At Driefontein, Diamond Hill, and elsewhere when the attack was continued till late in the day, the Boers seeing the attack closing in, and dreading the night charge, abandoned their positions.

Artillery and musketry preparation.
The hitherto supposed necessity of an artillery preparation to an attack is now proved to be fallacious. To shower shrapnel and high explosive shells on a position during the attack may be very necessary *to keep down the fire of the defence*, but it is quite incorrect to suppose that shelling a position before the attack commences does any material harm. At Belmont, also at Magersfontein, the positions to be attacked were well shelled the *day before*, but probably not the least harm resulted. The infantry behind cover need but "lie low" till the shelling has ceased, and the trenches being small and probably invisible are not likely to suffer much actual damage.

If the infantry and machine-guns can advance to within rifle range (and it is seldom that a position can be found so perfect that this is not to some extent the case), musketry will probably

THE ATTACK.

have an even greater effect than artillery firing shrapnel. This may sound a strange assertion, yet, it is clear that, if a shrapnel shell contains two hundred balls, a volley from two hundred rifles should have an equal or greater effect. But the riflemen could fire fully five times as many rounds per minute as the gun; so that forty riflemen may be taken to equal one gun at fairly close range. More shell are likely to go wrong—not burst properly, not fall at the required range, etc., than a corresponding number of rifle shots. Shrapnel bullets, it is true, may search trenches better owing to the greater angle of descent, but they have not the penetration of the rifle bullet.

Long-range artillery fire is not now of great practical use in keeping down the enemy's rifle fire. At a distance of 4000 or 5000 yards it would be very seldom that the estimated range was within 200 yards of the reality, yet not much harm would be done if the shell explodes 100 yards beyond or short of the desired spot. At shorter ranges it may of course be more effective.

Though the actual material results of artillery fire may not be very great, it has considerable moral influence, and will, undoubtedly, have great effect in lessening the fire and disturbing the aim of the defenders, so that it may be most useful when the attack is already well developed, especially for keeping down the artillery fire of the defence.

THE ATTACK.

Colonel Rimington, on several occasions, successfully attacked non-intrenched positions by shelling them, and at the same time sending a line of mounted men to gallop up to the position.

Infantry attack formation. Attacks over the open may then still have to be carried out, at all events as a demonstration to divert attention, or to advance against a position to ascertain whether or not the enemy be occupying it. This latter manœuvre was very often found to be necessary. At Poplar Grove, while the cavalry were making long detours round the flanks, it was quite uncertain whether the enemy were not still lying quiet in their trenches. The infantry, therefore, had to advance across a wide open plain in extended order (but found the position deserted). So also at Modder River, it was supposed that, if any, there were but very few of the enemy. But luckily and rightly the force was moved across the plain in extended order, else the result would have been disastrous.

It is interesting to note that even at the very commencement of the war, the attack formation as laid down in the older drill-books was discarded in favour of a more simple and practicable one.

That most usually adopted consisted of two simple and continuous extended lines, the rear one being perhaps 100 to 200 yards behind the front one. Such things as supports and local reserves were almost unknown.

In all tactical operations the cry must now be "extend." This not only applies to individual

THE ATTACK.

troops, but to formed bodies. We often hear of large masses of troops being kept together, especially in the Tugela operations, when a chance shell, perhaps not even aimed at them, has fallen among the crowd with disastrous effects. It is very evident that the old close formations are not suitable to modern warfare. It is all very well in a confined barrack square, but in the field, whenever there is a chance of troops coming under artillery fire, they should be extended.

Experience in the war has again and again shown, that what we have to fear, is not so much the individual aimed fire of marksmen, as the general hail of bullets fired in a given direction. Though rifles now carry further, the *targets* are no bigger than they were. To hit a man at an unknown range of over 1000 yards can be nothing more than chance. The man is but a speck ; the faulty aim of one shot cannot be corrected in the next, as on the ranges, and so many conditions, such as wind, light, etc., affect good shooting, that the best of marksmen could probably not hit a man once in twenty, or even once in a hundred shots, under the circumstances. Even at short ranges this applies to some extent. Yet if, instead of aiming at one individual man, you see twenty men together and aim in the middle of them, you are much more likely to bag one. So, if twenty men fire at the centre one of twenty others standing in a row opposite them, the chances are, several will be hit. If each aims at his particular *vis-à-vis*,

Reasons for extension.

there is every chance against his hitting him. So it may be said that there is every chance of a man being wounded accidentally and comparatively little of being struck by the bullet intended for him.

Therefore, if a man advances with no comrades within, say, twenty paces of him, he is not so likely to get hit as if he has a man a pace or two on each side of him. In order not to give any guide for aiming, invisibility will be one of the chief objects of attack formations.

If the firing line advanced shoulder to shoulder, though forming a by no means easy mark to hit at long range, yet it would be fairly conspicuous to see, and would form a target to aim at. If extended, and the wider the better, it would be practically invisible. The extent to which the individuals are to be separated will depend upon two circumstances. First, in order to give invisibility, the men should not be closer together than, say, two paces. Secondly, the amount of front which it is desirable to occupy will depend on the object to be gained. If it is only to "feel" for the enemy or to hold him, they may be very widely extended—forty or fifty paces apart.

On the other hand, we do not want the men so scattered as to be out of control, and, if a determined attack is to be made we require to have as many rifles as possible in the front line to be able to open fire whenever opportunity allows. So that then they should be as close as possible. The

THE ATTACK. 55

two-line formation has the advantage of being inconspicuous, yet readily closing to make an effective assault.

The formation in which a force will attack will be in one general line, with other supporting troops moving up a long way behind them. The exact formation of this front line is not of supreme importance. A hard and fast line, well dressed, and moving together, is not desirable. To maintain it thus distracts the attention of the officers and men, and unnecessarily fatigues them. Such a line is more easily discernible to the enemy, and if it should come under a flanking fire will suffer greatly. Even a charge of the enemy's cavalry is more likely to succeed against the flank of a rigid line than against an irregular string. It is better to move by extended groups in a general line, and if one group gets thirty or forty yards ahead, and another lags behind, it makes but little difference to the efficiency of the general line.

It has been suggested that a long advance in extended order is apt to cause loss of control and disintegration. But though in a very long advance it is possible that some portions of the line may be more delayed than others, this difference is not likely to be important. At Modder River three battalions of Guards moved in an extended line for some miles before coming into action, and though some portions had got several hundreds of yards ahead of others, the break was not very noticeable when the action commenced. At Poplar

Grove we advanced fully six miles in extended order, and presented a very fair line at the end of it. This, it is true, was in very open country.

The front line should contain as many rifles as possible when a real attack is intended. It may then be taken that the greatest force which can be hurled against a position at one moment (presuming the "reserves" must be some distance in the rear) would be in two lines with the men arranged two paces apart, or, *the attacking force cannot exceed one man per yard.*

The term "firing line" is now a misnomer, as probably the most telling fire will come from second lines and reserves firing over the heads of the front line, when opportunity allows.

Scouts out in front of the first line may be useful to give warning of ambuscades, etc., but they must not mask the fire.

Method of advance. The advance to the attack may be of long duration, and when close and effective range is arrived at, that is to say, "when the fire is becoming very effective," a different course will have to be pursued to that adopted when merely crossing the ground, and when, perhaps, not even under fire. It will then be desirable to adopt measures to prevent loss as far as possible.

The line may be broken into units, and the system of advance by short rushes of alternate units becomes necessary. As men cannot always crawl like lizards among the irregularities of the ground (though there are occasions on which this

THE ATTACK.

has been done), it is necessary for them to expose themselves while advancing; but the time of exposure may be very short if they can only move with sufficient rapidity. The soldier is only able to run rapidly for a very short distance, and therefore, as soon as he begins to slow down, it is better that he should no longer expose himself, but lie down out of sight till sufficiently rested to run hard again. Even if he run as long as possible at full speed, it may not be advisable for another reason. When approaching close quarters, the defenders will look out for a target, and as a man runs towards them, they will take aim and shoot him down. To aim thus would take, say, four seconds. If, then, each rush lasts only four seconds, the defenders will not have time to take accurate aim and fire. During the periods of rest, the men may occupy the time in shooting.

The "units" referred to should undoubtedly be as small as possible. Even sections are found to be somewhat unwieldy and large under heavy fire. Sub-sections or squads of eight to ten men are quite enough.

As regards the time to commence firing it will be dependent upon several conditions. As a great deal of ammunition may be expended in a long attack, and fresh supplies difficult to arrange, it will usually be best to husband the ammunition till decisive ranges are reached. Troops advancing, especially if over rough ground, are not easily seen at long range, and if they creep up without opening

Opening fire.

58 THE ATTACK.

fire till close, the attack will come with more of a surprise, and therefore be more likely to succeed. On the other hand, if it is desirable to make a demonstration of strength, fire may be opened at long ranges, and rapid firing kept up to simulate that of a large force.

Concentration of fire on one particular portion of the defence would be most desirable, but would not often be easy to arrange, except with the reserves and particular bodies. Men will naturally fire straight to their front.

Flank protection. In moving a line of troops across country where the enemy is known to be in force, great care must be taken lest he suddenly issues forth and makes a counter attack. If the flanks are not protected by cavalry or independent bodies, the flanks will have to be thrown back so as to be able to at once form a front to resist flank attack. Indeed, a diamond-shaped formation has been suggested as the proper one to adopt.

Supports. The use of long-range weapons causes the modern battle-field to be greatly enlarged.

In olden days, with a line of troops drawn up within a few hundred yards of an enemy similarly disposed, if a gap of a dozen yards occurred it would cause a weak spot in the defence, or render the assault not so homogeneous and complete, so that a local reinforcement would be necessary to fill the gap. It is a very different matter when the lines are extended and spread over miles of country, and when the critical stage of the battle

THE ATTACK.

is usually over before the attack has approached within 400 or 500 yards of the defensive position. Then a gap of a few absent men makes but little difference to the general line.

The replenishment of the whole line—that is to say, the maintenance of effective fire—is another matter.

With the curved trajectories of old rifles, supports could follow fairly close in rear of the firing line, without fear of any shots passing through the latter striking them. With the flat trajectories of modern weapons a bullet aimed at a man of the firing line will, if it passes him, travel to a considerable distance behind, and, moreover, the penetration being so much greater, it may even do damage in rear after having passed through a man in front. This would necessitate the supports being much further in rear than formerly, especially at close quarters, when the trajectory is nearly flat. That, however, is just the time when supports should be close up to be of any use.

Altogether, then, there is no advantage in having supports close in rear. They may just as well be *in the firing line* (except the latter be so thick as to become more conspicuous). Then a greater proportion of rifles can be utilized when required.

If the troops in the front line be checked at any point owing to the deadliness of the fire poured into them, it is scarcely likely that any advance can be better made by bringing up reinforcements into that line. It would merely result in a greater loss of men, since the target would be larger. If the same number of enemy continue firing, the zone will be none the easier to cross because there are more men to do it. If at a long distance from the enemy, the moral effect of numbers will not be so great a factor as, for instance, in storming a position. It is not as though the men were dummies, and if a certain number are hit the rest will go on. Once men begin falling thick and fast, their companions will hesitate to proceed, and leaders, if they be wise, will not discourage such caution.

Reserves. But though *local* reserves may not be necessary, it is of course desirable always to keep a *general* reserve in hand.

Such a reserve kept well back out of fire may be used in many useful ways; not only for replenishing the fighting line, but for extending the flanks, conducting a flank attack, guarding against counter attacks on flanks and rear, or for taking up a suitable position whence a steady long-range fire can be kept upon the enemy. Troops in reserve would also be of great value in case of a counter attack by the enemy, when, if the front line was driven back, they might take up a position whence they could prevent pursuit, and enable the first line to re-form behind it.

A reserve may also be desirable in attack formations as a *moral* support. It gives confidence to a man to know that his friends are behind him "backing him up."

THE ATTACK.

It is very necessary to keep a continuous hail of fire playing on the position to be attacked, to keep down the fire of the defence. But it is necessary to use caution, and be sure of not firing into the backs of the firing line, who, in inconspicuous uniform and among rocks or bushes, may be practically invisible for a time.

At Belmont, the second line coming on a considerable distance in rear of the first, presumed, for some unexplained reason, that the kopjes in front had not been carried, and that the first line was checked at the bottom, so opened fire on the kopje, which was, however, covered with our men!

We are, perhaps, apt to forget that the trajectory of the rifle will generally allow of troops in rear firing over the heads of those in front, on level ground. Remember that with 1000 yards' range, the bullet will pass 25 feet above the line of sight 500 yards off.

The distance which the reserve must be behind the first line, so as not to come under the same fire, will be *at least* 300 to 400 yards, but it would usually be better much further back.

As regards the strength of the reserve—that is, the proportion of a force which it is desirable to keep back—much will depend on circumstances. If the extent of the ground on which the advance has to be made is confined and limited, there may not be room to extend all the men of the front line at a sufficient interval to be inconspicuous. The surplus would then have to be sent back to

62 THE ATTACK.

the reserve. If the ground was very open, it may be desirable to put as many men as possible into the front line.

In skirmishing, or feeling for the enemy, only as many men as will make a show need be in front, so that the rest may be more in hand in reserve. On the other hand, in such reconnaissances, it will generally be advisable to cover as great an extent of ground as possible, so that all the men available may be used. I have known 200 men to advance on a front of fully three miles.

But it is, of course, always desirable to have *some* body of men in reserve in case of emergency, and when there is a probability of a fight, the larger it is the better.

As an extended line is the best, or, indeed, the *only* formation in which to move troops under fire, the reserve becomes practically a second line.

Final stage. When the position has been closely approached, the first line checked, perhaps, by the deadliness of fire, and reserves brought up to endeavour to subdue it, we have got to a stage in the attack in which another change in the old order of things has come about. The time-honoured charge with the bayonet is not likely to be a great feature in the wars of the future. If the attacking force is able to advance to within 300 or 400 yards of the defenders, it will almost always be the case that either the latter will abandon the position and fly at such a rate that bayonets will not harm them, or they will stand and be able to pour such a telling

THE ATTACK. 63

fire into an assaulting line as to hurl it back. While the assailants are crossing this zone, the defenders could fire fully 15 or 20 rounds a-piece.

Under these circumstances it will almost always be better for the assailants, if the defenders remain at their posts, not to advance under such a deadly fire, but to take up the best positions they can and fire away steadily, to subdue, if possible, by fire alone. If the defenders are not then very well posted, they will suffer as much as the attackers, and as the latter will presumably be in greater force they should win the day.¹ Boer attacks usually culminated in this way.

But though the bullet may be the chief factor of modern warfare, the bayonet may on occasion still count for something. Men's nerves are still delicate, and when a practically beaten or demoralized body of men see a sudden rush upon them of glittering steel, it must have a very great moral effect. We often heard how the Boers dreaded the lances and bayonets, and how on one occasion (a sortie from Ladysmith) the very cry of " Fix bayonets!" caused them to fly in terror, even though there were, on that occasion, no bayonets to fix! *Charging.*

If the attack fails, to retire steadily under fire is a difficult and trying operation. Once the word is given men like to jump up and run back. But this gives an opportunity to the enemy to pour in a hot fire. Most of our losses at Magersfontein are said to have been during the retirements. *Retirements.*

64   THE ATTACK.

They should, if possible, be carried out similarly to the advance—that is, by short rushes and in small lots.

As orders to "retire" may easily emanate from irresponsible men, undesirous of pushing on, so subordinate officers, anxious to encourage their men forward, may question the validity of the word to retire, when passed from mouth to mouth down the line. I remember at Modder River this occurred, and that we anxiously shouted back, "Who gave the order to retire?" though no answer was forthcoming.

Still, it does not do for an officer in such circumstances to take upon himself the countermanding of it. At Stormberg, when the order to retreat was given, several officers, doubting its being genuine, prevented their men falling back, which only added to the disastrous result.

Other methods of attack. But in speaking of attacking, it has almost become implied that we mean advancing across an open fire-swept zone. Yet much will depend upon the ground manœuvred over, and, as often as not, a great part of the advance towards the position may be more or less under cover, and advantage may be taken of folds of ground, and of rough country with sheltered nooks.

Different formations may be more suitable over undulating and broken country, where cover is plentiful, where second lines may line the hilltops and keep up a hot fire, while the front line advances across the intervening valleys. So too, masses of troops may have to be squeezed through intricate defiles, through streets in a town, or along roads through woods.

A form of attack, which may be called the Boer method, is by sending only a few men to creep from rock to rock, or bush to bush, crawling on all fours over the open, putting in a shot or two whenever they get the chance, and using their own discretion of pushing forward. Gradually they *converge* on the prearranged point, while others from behind press on individually to support them.

Such an attack may often succeed against a force badly posted, even though superior in numbers. The capture of Majuba Hill, in years gone by, is a well-known typical instance.

In advancing to attack through woods and bush (such as we had in the bush-veldt of the Northern Transvaal) several precautions must be taken. The advance must be very slow, and great care taken not to lose touch. Movements then are extremely difficult to control, and, unless a very steady advance is made, the force is liable to get scattered. Small units must be kept well in hand. **Attack in woods.**

When the advancing troops become heavily fired upon, the best, if not the only way, is to open a heavy, promiscuous fire back (even though no target is visible), especially using machine-guns and artillery to thoroughly search the woods in front. This generally had the effect of soon clearing the front. In the advance to Mafeking the enemy was encountered in a wood. As soon as

THE ATTACK.

we brought our guns and pompoms into play, though no enemy was visible, their firing soon ceased.

In the northward move to Pienaars River through the bush country, when a few snipers opposed our advance, it was usual to "sprinkle" the front with machine-gun fire, which greatly facilitated reconnaissance.

Attacking a force on the move. One of the best opportunities for successfully attacking an independent force is whilst it is on the line of march. It is then bound to fight wherever it happens at the moment to be, and thus, not only is no choice of ground allowed to the defender, but the attack may, and should, be delivered at the particular moment when the enemy is crossing a tactically bad piece of country. This would usually be where no natural cover was available for the defenders, while the attackers could utilize ground favourable for their approach.

If information be brought that the force is advancing, or is likely to advance along a given road, its destination can generally be surmised, and then a certain length of the route is open to the choice of the attacker.

The attack under such circumstances is not denied any of its usual advantages, such as option of the point of assault and choice of time. In addition, it gains the advantage of having the selection of ground to advance over, knowledge of ranges, and ability to closely reconnoitre, just beforehand, the position to be occupied by the enemy. The defenders, on the other hand, are

THE ATTACK.

deprived of most of the advantages inherent to defensive tactics. They are unacquainted with the ground, have but a poor choice of position, are unable to arrange beforehand the exact disposition of the various units, are crippled in their arrangements for supplying ammunition, reinforcing, etc., and find themselves attacked when in a formation probably not suitable for defence. In all probability they will have no supply of water, and would, on this account, not be able to hold out through a protracted engagement.

Attack under such circumstances should then be very promising of success.

There are three conditions under which such an attack may be carried out. Either (1) opposing the moving force in front; (2) attacking it on the flank; (3) pursuing in rear. All these may be combined to some extent.

The first has this drawback, that if there be an efficient advance guard well to the front, the attack may be stayed until the force is formed up and ready to resist. The attack in such cases must be very determined and sudden.

Attacking on the flank, especially if a feint be made at the same time on the other flank, would always have the greatest promise of success. It would require careful timing, but the column could always be delayed by a small force obstructing the road. The attacking forces would be drawn up at such a distance from the road, and so hidden as not to be seen by the advanced guard or flankers.

68 THE ATTACK.

Attacking a column in rear is not so likely to prove successful, since it may proceed on its way fighting a rear-guard action. If the road could be obstructed beforehand, or a small mobile force sent round to "head it off," a sufficient delay may be caused.

On several occasions we have pursued a Boer force, but owing to their marvellous mobility, it nearly always resulted in a rear-guard action. By their good information and knowledge of the country, we were seldom able to intercept a commando on the march.

The Boers several times attacked our columns on the move. The disaster to the Dewetsdorp garrison retreating on Reddersburg is a good example of a force on the move having to take up a bad position when attacked. Colonel Benson's column, too, at Brackenlaagte suffered very severely by being attacked when on the move. Other instances are the attacks on Lord Methuen's force at Klipdrift, and on Colonel Von Donnop's convoy, near Klerksdorp. There were, it is true, several minor cases of convoys being attacked unsuccessfully by Boers; but, in most of these, the tactical advantages were missed, the enemy advancing to attack over open country, or when the moving column was on ground affording good cover.

Attacking camp. In attacking a camp, when the position is not naturally good (and how often in practice are tactical advantages sacrificed to convenience !) and

THE ATTACK.

the outposts are close in, or weakly posted, the whole force may be partially surrounded by the assailants, who, driving in the outposts or approaching as close as practicable, may open a concentrated fire on the camp with such effect as to force a retreat and abandonment of the position. This should be specially likely to succeed at night. Something of the sort happened to General Clements at Nooitgedacht, who found the fire from commanding heights (whence his outposts had been driven in) so effective as to compel him to retire and abandon his camp and stores.

Although, then, a successful frontal attack may occasionally be delivered, it will in most cases undoubtedly be preferable to work round the flanks and cut off the force from retreat, presuming, of course, that the attackers are in superior force. If this had been attempted more often in the earlier part of the war, we should certainly have had more decisive results. But the Boers were especially apprehensive of this course of action, and always retired as soon as their flanks became threatened.

Surrounding a force.

It does not, however, follow that, having gained the flanks, or even rear of a force, it is necessarily at a disadvantage, but if the attacking force be preponderating, and the defenders either in a bad position or short of supplies, it will usually be so.

It has already been said that *strategical* progress can be accomplished without necessarily making tactical attacks. So also a tactical advance may be made, avoiding attack and acting on the defensive

Strategical advance. Tactical defence.

when necessary. There are plenty of minor instances of this in the recent war. In the advance to Mafeking, for instance, it was ascertained that the enemy had taken up a strong position among some hills near Maritzani through which our road ran. It was at once decided to turn off and move by another road running some three or four miles to the westward. The enemy, seeing this, had no alternative, in attempting to stop us, than to move across to attack us on the flank while we stood to receive them. Thus they were forced to *attack* instead of to *defend*, and were easily driven off. The engagement on the Molopo River a few days later may also be cited as an instance. Our whole force advanced (in a large circular formation with transport and guns inside), and when the Boers attacked we halted, and stood our ground on the defensive. The first attack being repulsed, we moved on, but halted again as subsequent attacks were made.

This is, after all, no new principle. It is but that of the old hollow square formation of the Soudan campaigns, only that the individual troops are now so greatly extended that the square becomes hundreds if not thousands of yards across.

Hurried movements undesirable. Though rapidity of advance is of the greatest importance when the enemy is to be taken by surprise, hurried movements are not always desirable. At Modder River, though the situation may have appeared a very awkward one for us during most of the day, when almost the whole of the British forces were lying flat on the ground, unable to move, yet the Boers were equally unable to advance from their trenches, and even any move towards the flanks such as they attempted, had to be conducted over the open and met with no success, since they came under our fire. It was, therefore, a position of "stalemate," except that our artillery was so much the more powerful that it was bound to tell in the long run. Had we tried more active measures with our infantry, it would certainly have resulted in greater loss and, at most, only hastened the inevitable end.

Mounted troops may sometimes be able to gallop across an open, fire-swept piece of ground, and do it so quickly as not to allow the defenders time to pour in any great amount of fire. In this way they may often approach close to a position, and, if the latter be on a hill, may get into dead ground at its foot, when they can dismount and creep up to close quarters. *Attack by mounted troops.*

The attack by mounted, or rather dismounted, men requires careful consideration, as it was an event of daily occurrence in the late war.

Mounted men will frequently have to attack positions. It may be to drive away a few men of a rear guard hanging on to a hill. It may be to reconnoitre to find out if a position is held in strength, or it may be even to attack a regular position which it may be of the greatest importance to obtain immediate possession of.

Horses must, as a rule, be left as soon as the fire proves destructive. It is usual with us for one man to remain behind, leading three other horses. This is a good system, as the man is then able to take the horses back under cover, and may bring them up when desirable, but, on the other hand, it only enables three men out of four to proceed with the attack.

The Boers had their horses so trained that when they dismounted, the horses stood where they were till fetched. This enabled all the riflemen to move forward, but the horses had to be left under cover, which might necessitate a longer tramp for the men, and they had to return back to the horses. Also the animals were liable to be stampeded by a shell, etc.

If the enemy retire, or any means is seen of doing so, the horse-holders should bring along the horses and follow as closely as is safe the firing line; while the dismounted men should be ready, directly they see an opportunity, to mount their horses and pursue, or move to a flank or elsewhere.

Care should be taken not to mass the horses too closely. Many instances occurred where a crowd of horses taking cover have been viewed from afar, and a few shells promptly sent among it.

Fortification in attack. Intrenching is often considered only of use in defence, but it is no novelty to utilize it in the attack. I saw it tried years ago in manœuvres in Germany. At Modder River, when the enemy's fire came very hot, we scraped hollows in the sandy ground with our bayonets as we lay. At Magersfontein, more elaborate cover was made in

THE ATTACK.

some of our advanced positions, and it might have proved beneficial if these had been improved and held on to instead of being withdrawn from.

At Paardeberg the Canadians and the Gordons advanced under cover of dark with bags of sand. Immediately their presence was detected and fire opened upon them, every one lay down and did what he could to erect cover.

In darkness or fog an advance may be made close up to the defensive line and a good intrenched position taken up.

It is worth while bearing in mind several small details which have their effect, however trivial, on the result.

Details.

In advancing to attack, artificial obstacles may be encountered. Most military movements have their converse, which should be studied. In considering how to defend a place it is well to think how it is likely to be attacked. So, to overcome obstacles, is but part of the subject of how to construct them. This will be treated of in another chapter (*vide* p. 196). But it is always advisable to remember the maxim, "when in doubt, lie down," which is especially applicable in this case. The obstacle can be surveyed, it may be crawled under, or it may be cut or pushed down while one is lying in comparative safety from fire.

It was a practice of the Boers, when about to deliver an attack at night, or reconnoitre a position, to drive some loose horses towards the lines. Should the sentries not be alert, advantage may be taken of that fact. If they challenge or fire, their exact whereabouts is known. The existence of wire entanglements, mines, etc., can also be thus ascertained, and the obstacle destroyed.

On approaching the trenches at Magersfontein (after the evacuation) a number of old horses were driven before us to explode any mines, etc., there might be.

Caution must always be used when attacking to prevent being ambushed. An enemy may purposely evacuate one position only to fall back on a stronger one behind, and, if the assailants, having charged up to one line of defence and finding it abandoned, rush onward, exhausted and dispersed, they will suffer terribly if suddenly confronted with a second position.

At Buffelsfontein the Imperial Light Horse galloped up a hill on which no enemy was visible. But suddenly a terrible fire was opened on them at close quarters by the well-hidden foe.

It is not advisable to attack eastwards about sunrise, or westwards about sunset; else the sun in the eyes will be apt to act detrimentally to the shooting.

Attacking in fog is likely to result in mistakes and losing direction.

Burning the grass, or bush, especially if the wind be towards the enemy, may often cause a screen to advance under. Far-away movements, retreats, bringing up reinforcements, etc., may also be hidden by this means.

It is desirable to consider beforehand what will be done with the position when taken. It will usually have to be defended against a possible counter attack. If so, is it suitable for defence, and are the means at hand to intrench it? At Spion Kop, after capturing the hill, it was found to be very difficult to dig trenches; the ground in front was so little exposed to fire that the enemy could creep close up, and there was no good water supply available.

CHAPTER III.

DEFENCE.

Advantages of defensive tactics. It has already been said more than once, that defensive operations have increased in importance by the introduction of improved firearms, and the reasons for this are explained in the last chapter. In general, they are that a man lying behind good cover, and so rendering himself more or less invulnerable, is now able to strike at his enemy at a further distance, and deliver a greater number of missiles in a given time than formerly. If he can hide himself away, his position is not now disclosed by the smoky discharge of his rifle. And the same conditions apply to a gunner with his gun. Moreover, formerly it was possible to reconnoitre a position and ascertain its strong and its weak points. Now, with smokeless powder and scattered forces, it is extremely difficult to find out exactly "how the land lies," and the lack of such knowledge adds another great difficulty to the attacker.

It never does, however, to be too theoretical in war. A badly chosen position, faulty dispositions, lack of vigilance, or insufficiency of

DEFENCE.

ammunition, may render all else abortive. Even apparently good British troops, supposed to be carefully posted, have been driven from their position by no such greatly preponderating force.

The power of defence is chiefly dependent on two factors: (1) the suitability of the position; (2) amount of the supplies. If the latter be limited, it may be considered that the position includes the line of communication which will have to be guarded and kept open.

The first and most important consideration in taking up a position for defence, is to have a *clear field of fire*. It is important that such a space exists all round, so that flank or rear attacks will not come unawares. If the enemy can creep up among bushes, or rocks and undulations, he can approach close without loss and without being seen, and is then on almost an equal footing to the defender. **Position for defence. Field of fire.**

The second consideration is *cover*. This is sometimes considered first, but it must always be remembered that a man lying down, even in the openest ground, is but a small mark, practically invisible over 500 yards. If the ground be very open, the attacker is much more conspicuous, especially while moving and exposing his whole body as a target. **Cover.**

Therefore cover is not so important as field of fire.

But cover is, nevertheless, a very important consideration, and it is very seldom that it is to

78 DEFENCE.

be found all along a large position. And cover is not only of value for the actual protection against bullets, but its moral influence is very great. Men lying in the open, or behind indifferent cover are, when exposed to a heavy fire, very apt to keep their heads flat on the ground, and will not raise them to fire back. Really satisfactory natural cover is seldom to be found, and it is only on positions carefully chosen at leisure, that such can be relied on. For this reason, artificial intrenchments are of the greatest advantage.

Given three conditions: (1) A position with clear field of fire for at least 500 yards all round; (2) Plentiful supplies of ammunition, water, and food; and (3) trenches or walls with head cover, and, so long as the outposts are alert to give warning, it seems impossible for an attack to succeed. Be the numbers ever so predominating it matters not, for the more opponents there are, the thicker they must be, and consequently more vulnerable to the hail of fire pouring from the trenches. It would be but a very few chance shots from the attack that would penetrate good loop-holes; and while the fire lasted, no close advance could be made.

Ability of small posts to hold out. Reliance on the ability of a small fortified post to hold out against great odds is a point on which some people disagree. They will say such weak garrisons are liable to be "mopped up" in detail. While there are, however, but a very few examples in the recent war of small well-posted garrisons surrendering (and these do not really affect the principle, since grave faults were in each instance undoubtedly the cause of the defeat), there are many of their holding out against great odds.

Of the former, the following are the only instances I know of:—

Modderfontein, some 200 troops surrendered to 2000 Boers.

Klerksdorp, surrendered without a shot being fired.

Helvetia, said to have had but very feeble defences made, and the outposts were presumably anything but vigilant.

Kuruman (second investment, Jan., 1900), only a few local police and volunteers. Held out till guns brought to bear.

Jamestown (June 2, 1901), defended only by local guard. Post on hill surprised and taken. Remaining works held by forty men for three hours *without suffering a single casualty*, and then surrendered. Enemy said to number 1500, and to have lost twenty-six men shot.

Blockhouse, near Brandfort (Aug., 1901), owing to defective construction, bullets penetrated walls. Sergeant in command killed.

Also some instances of works being rushed *at night* (Belfast, etc.). And a few blockhouses with half a dozen men surrendered.

Against these, may be quoted the following instances of small garrisons successfully withstanding attack by large numbers (besides the larger operations of Mafeking, Kimberley, etc.):—

DEFENCE.

Wepener (thirteen days, 1450 and 7 guns. 5000 to 6000 Boers ; 9 guns).

Elands River (400 and 1 gun *v.* 2500 Boers and 6 guns).

Kuruman (Nov., 1899. 75 men held out for a week).

Lichtenberg (600 *v.* 1500 Boers).

Winburg (250 and 1 gun *v.* 1000 Boers and 2 guns).

Itala Fort (220 men, 2 guns. 1800 to 2000 Boers fought for 19 hours).

Laybrand (150 men for 3 days *v.* large force with several guns).

Fort Prospect.

Kaalfontein (120 men held out for six hours).

Zuurfontein.

Phillipolis (40 civilians held out for a week *v.* 600 Boers).

Fauresmith (Oct., 1900).

Sand River Bridge (June, 1900).

O'okiep.

Various blockhouses and posts on the railway.

General disposition of force for defence. Long extended lines will usually be necessary to prevent the enemy from outflanking the line of defence. Yet the actual length of extension between individuals and groups will greatly depend on the strength of the position, and the numerical strength of the enemy.

Long-range weapons enable a line to be more widely extended, for with rifles only capable of doing damage at 200 yards, every 200 yards of the position would require careful guarding. Now, if a gap of 600 or 800 yards be left, or lightly held, no adversary could advance through it without coming under a heavy cross fire from each side.

For this reason, if the position be not over a mile or two long, it may be best to put almost the whole strength at the two flanks, holding the intermediate part very weakly. An attack on the centre would then be exposed to a cross fire from both flanks; while an attack on either flank would be met by a strong opposition.

If, in a big position, a naturally weak spot occurs, it may be advisable to purposely leave it unguarded. The enemy may break through; but, the contingency being anticipated, he may then find himself in a very awkward position. This was practically the case at Mafeking, when Eloff and his party got through the outer defences, but were then forced to surrender.

It will always be desirable in defensive operations to deceive the enemy as much as possible. The less he can discover the positions of the works, the whereabouts of the guns, and the strength of the garrison, the better. At Colesberg, squadrons of cavalry were sent out at night, and marched in, in full view of the Boers, in the early morning, giving the impression of the arrival of reinforcements.

False works rapidly erected, a row of hats on sticks, some cartridges thrown among the embers of a fire, may all serve to mislead the enemy, and attract his fire away from more vulnerable points.

F

Concentration of garrison undesirable.

It is most fatal in defensive operations to crowd the force into a small space. Directly men get together they form a target, and when a concentrated fire is brought on them, whether only musketry or artillery, the casualties are sure to be great. The mutual support derived from being close together is of no avail against bullets; quite the contrary. Even though in trenches, or behind good cover, this still applies to some extent.

The reason of this is worth careful consideration. Take two examples:—First, suppose a position to be taken up on the hill A, the men being placed shoulder to shoulder. The reports of the rifles, the thin film of smoke from so many discharges, the dust raised, the occasional showing of a man, will soon disclose the position to the attackers. The enemy attacking from all around the front will pour in a terrible concentrated fire on to the position, and however good the cover, many shots are bound to tell. The enemy, constantly watching the one spot, will frequently get an opportunity for a good shot. But there is more than this. A bullet that misses the man it is aimed at, may easily hit one on the right or left. Hundreds of shots may be fired at the position, and this hail of bullets will probably have effect on some of the defenders. A lucky shell may have a very big effect.

But now take the second case. Let the same force be spread out along the ridge from B to C and presume its flanks are secure. A very different sort of target will now be presented—in fact, it is practically as large and disseminated as that presented by the attackers. Put each man behind what cover he can get (and natural cover thus be

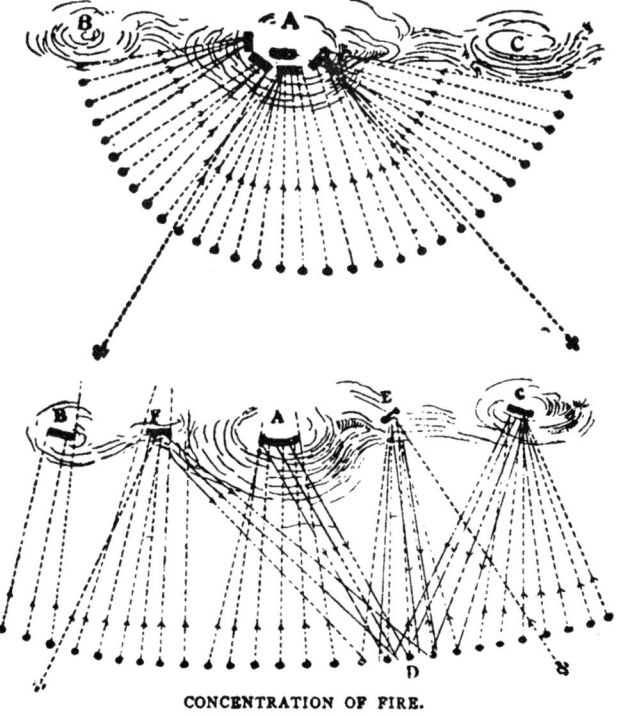

CONCENTRATION OF FIRE.

more easily taken advantage of). Now the enemy hear a report, first in one place, then in another. No idea can be got of the exact line of the position, or whether the defenders are in two or three lots, or whether in an extended line. They can only fire on the general line. There will be no collection of smoke to guide them, and when watching one part of the line, they may not notice a man expose himself at another point.

Then, if they fire at one individual and miss him, it is very unlikely that the shot will be so much to one side as to strike the next man. The hail of bullets will be extended over 400 yards of country, so that a very large proportion of them will pass between; nowhere near any defender. Artillery fire is not likely to do much harm.

If a force be suddenly attacked, when not in a good position, then, more than ever, is it necessary to prevent the natural tendency of men to huddle together. Concentrated in a small space without good cover, they are at the mercy of an encircling force bringing a concentrated fire to bear on them, while the attackers are distributed round, offering only a widely extended target. If the defending force, however, at once extends, it will be able to avail itself better of any such cover as does exist, and, as regards target, will be no worse than the assailants; and probably better, since they can lie low to defend, while the enemy must expose himself to advance.

At Nicholson's Nek, the official despatches say, " The advanced parties were driven back on the main body in the centre of the plateau, and the Boers gained the crest-line of the hill, whence they brought a converging fire to bear from all sides on our men crowded together in the centre, causing much loss."

Other instances could also be quoted where a force, being pressed, had closed in, and gradually become a compact target, and suffered accordingly. There are instances, too, of the reverse proving successful. A good example is the action on the Molopo, when the Mafeking relief force, while on the move, found itself completely surrounded. The advanced, rear, and flanking guards, which were rapidly reinforced, formed roughly a large circle a mile or more across. The circumference was manned by under 1800 men, so the line of defence, including reserves, would have averaged only about one man to every three yards. Though shell after shell was thrown into this great area from all sides, but little harm was done, as there were so few men inside to catch the fire; and the result was the enemy, probably greatly outnumbering us, were not even able to stay the advance of the force.

It will always be desirable, when an attack is expected or is likely, to allot each man to the particular spot he is to occupy; whether it be for cover behind a particular rock or heap, or to fire through a certain loophole in a trench. Then if a sudden alarm is raised each man knows where to go, instead of two or three men trying to thrust their rifles through the same loophole or crowding together behind an inadequate bit of cover. *Allotment of troops.*

As regards the number of men per yard necessary for the efficient defence of a given position, the following considerations will have to be taken into account :—

Number of men necessary.

1. The *enemy's* strength and armament.
2. Whether the *position* is a circular one; such as that surrounding a town, where reinforcements from the centre can be rapidly sent to any desired point. Or if it form a long line, and if so, whether means, such as a railway, exists to despatch reinforcements by.
3. Nature of the *ground* and *artificial improvements*, i.e. if good cover obtainable, if approaches open, if rapidly entrenched or careful fortifications made.

It will usually be best to spread the force out; both so as to form a bad target, and to prevent, as far as possible, outflanking movements, or to enclose a large area of ground. But mutual support must never be lost sight of.

For many reasons, too, it is necessary to keep men, to a certain extent, together. They must be under the control of one head; the proximity of comrades gives mutual confidence; and the supply of ammunition and provisions is easier worked. A number of very small posts would usually be better than a few large ones. A section of twenty to thirty men has been found to be the largest suitable unit for command in action. More than this form an extended line too great for one commander to properly control in action. A group or sub-section of eight to twelve men is preferable. Such detached parties may well be thirty to forty yards apart if necessary.

The greatest strength necessary to defend a position must depend on the greatest strength that can be brought against it. Now we have seen how the attacking force would seldom be in closer formation than that of a front line with intervals of about four paces between men, backed by two or three other such lines. So that it is unlikely that the attacking force would exceed at one time one man per yard of front, or, say, 2000 *to the mile.*

The question of the relative strength of the defending force to ward off attack is difficult to decide, considering how greatly it will vary with circumstances; but practical experience seems to show that troops *well intrenched* can certainly keep off a force of eight to ten times their number.

This would imply that 200 to 250 men per mile should be usually sufficient to hold, at all events temporarily, a good intrenched position.

But if the position be not very favourable, or has not been artificially improved, probably fully two or three times this number will be necessary.

At Wepener the perimeter of the defences was over seven miles long. The garrison distributed in them would run about 200 to the mile. At Mafeking the proportion was only about 100 to the mile.

DEFENCE.

If the position is such that the enemy can attack on all sides, *i.e.* a circular position, their line of front will, of course, be longer than the concentric inner line of defence. They can then bring a larger proportion of men into action without unnecessary crowding.

It is impossible to quote many instances in the South African War of the relative numbers of forces acting on the offensive and defensive, since the exact strength of the Boers is but seldom more than guess-work; and their independent methods in attack render it impossible to say how many really pushed an assault home and how many hung back under cover till all was over. The few instances where fairly approximate estimates can be made are as follows:—

At Paardeberg we know that 4200 indifferent, half-starved Boers, intrenched, but without much head-cover, in good position, with six guns, repulsed about 16,000 good, disciplined, fresh British troops, backed by over four batteries of artillery; this would give a proportion of just about four to one. At Modder River the Boers were said to number about 3500 against our 8000. At Itala the garrison of a little over 200 repulsed the Boers said to number from 1800 to 2000.

But though we may speak about the *strength* of the defence and of the attack, it must not be forgotten that nowadays strength implies *amount of fire*, rather than number of men. If one side is armed with a specially rapid firing weapon,

J. E. Middlemore] HART'S HILL, SHOWING THE SANGARS. *[Photo, Durban.*

much fewer men will have the same result as a larger force with a slow feeding rifle.

When a mixed force takes up a long defensive line, mounted troops should form the reserves, able to rapidly move to any threatened point to reinforce it.

In calculating the number of men necessary to garrison a position, allowance must be made for casualties. These will depend on the efficiency of the cover. With really good intrenchments the losses should be very slight.

Sickness will also have to be taken into account. This will vary according to the climate, the healthiness of the locality, and the food and water supplies. In Ladysmith, where the health was not good, thirteen per cent of the garrison was at one time on the sick list.

As a general rule, a semi-permanent position will either be circular, such as that enclosing a town or camp, or a more or less straight line. The former enables a thin line of outposts to be rapidly reinforced from the centre, and will have the advantage of guarding a considerable area of country for grazing, etc., and renders the supplying of the line comparatively easy. On the other hand, when attacked, if the circle be small, projectiles from the enemy going high would go towards the camp or town; a converging attack can be brought upon it, and portions of the position enfiladed, while only about a quarter of the line can fire in our direction. So that for a *small*

Semi-permanent position.

position it may not be so favourable as a line, provided the flanks of the latter can rest on a secure obstacle, or are strongly held. A semi-circular position is a go-between, both as regards advantages and objections. In the selection of a position everything will, however, depend on the natural features and objects.

Distribution of garrison. One of the most important questions to be decided before occupying a long position for defence, is whether to distribute the whole available force equally along the line, to weakly hold the front with outposts, while keeping the remainder in reserve ready to move up when and where it be considered desirable, or whether to distribute the garrison in a chain of detached posts. There is much to be said in favour of each system. It is always preferable to have the men in the actual positions they are to occupy, ready for all emergencies. They should certainly be so posted when attack is thought to be likely. As the attacker has the choice of time and point of attack, the defenders should always be as strong as possible at every point, without wasting their forces by keeping a large proportion in reserve. For instance, Cronje with his 8000 men at Magersfontein could occupy his trenches in such strength (nearly 1000 men to the mile) as not to need any reserves.

It was very generally the custom in South Africa to house the men in, or close to, the works they were to occupy. The available troops, however, may not be sufficient to strongly occupy the entire

J. E. Middlemore] CONTINUOUS TRENCHES ON VAAL KRANTZ. [Photo, Durban.

length of the line. Then, of course, reserves will be a necessity, to be moved to wherever the attack appears to be developing. But these must take time to assemble and to march to their destinations, and time is often of the greatest value. Yet reserves can, so to speak, be formed out of the firing line by taking a certain proportion from each unit, and, by closing them in towards the threatened point, a concentration may be formed more quickly than could be done by moving up reserves from a distance.

In a long line, however, the distribution of supplies, especially water, may be a great difficulty, and it is often for many reasons considered desirable to keep men together in one camp as far as possible.

The Boers at Magersfontein held an exceptional position, for the trenches were constructed close in front of a long ridge of kopjes, behind which extended a straggling camp practically safe from fire. So the men in camp could, at a few minutes' notice, line the trenches, in which, however, a number of men were apparently always on duty.

When very long lines of defence have to be taken up, such as the strategical lines referred to before, numerous small, independent observation posts, well entrenched, with reserves at intervals in rear, would seem the best system. These lines may be very attenuated. Groups of half a dozen men at wide intervals, perhaps as far as half a mile apart, may stop an advancing force for some time, *Long lines of defence.*

while reserves are being brought up to oppose it. Frequently our columns in South Africa were delayed for hours by a few sniping Boers, occupying good positions whence they were not easily dislodged, because of the uncertainty as to whether a much larger body was not in position awaiting our advance. If reserves were posted every six miles, they would always be within two or three miles of any point of the line, and if the attempt to break through was as far as possible from their reserves—that is, midway between two—those from both sides could come up to oppose it within an hour.

Many long lines were occupied by our troops in clearing up the country towards the end of the war, but they were usually posted in block-houses, which were so conspicuous and so limited in size, that the enemy could always see exactly what force they had opposed to them.

To be secure against individuals and small parties slipping through by night, it will be necessary to have the posts not more than 400 yards apart, and each would have to furnish a sentry. Six men under a non-commissioned officer was the usual garrison of the block-houses.

These posts should be arranged, as far as possible, so that in the event of their opening fire, they will not shoot into one another. This is often not so difficult as it may at first seem, as, for instance, by posting them on opposite slopes of a hill, or arranging that a small mound, or even a wall, intervenes.

DEFENCE. 93

At Lake Chrissie, during a night attack by the Boers, two portions of our force fired into one another with disastrous effect.

All this refers chiefly to fortified positions for more or less permanent defence. It will frequently occur during active operations that one or other force will, for a time, act on the defensive, wherever it may happen to be. This will involve very different conditions to those appertaining to the deliberate occupation of a well-chosen position. Everything will then be dependent on the natural features of the ground. Occasionally a site may

Temporary defence.

POSTS PROTECTED FROM EACH OTHER'S FIRE.

be found which is nearly as good, as regards open field of fire, cover, protected area, etc., as a well-prepared position, but much more usually it will have its weak points, covered approaches, lack of cover, and other unfavourable features.

Then, as a rule, the force holding it cannot be so spread out without risk of the enemy breaking in, so that a smaller area will have to be occupied (which can perhaps be extended as time goes on). Every rifle must be ready to open fire, since there is not likely to be much time for moving up of reserves. But this does not necessarily imply that all the men must be in the most advanced line.

94   DEFENCE.

Many may be posted in commanding positions hundreds of yards behind the front troops.

Advanced posts. Small detached posts, well in advance of the main position (and protected from its fire), will often be of use in giving warning of attack, in keeping off hostile scouts, in puzzling the enemy in endeavouring to ascertain the limits of the position, in preventing a near approach of the enemy's guns, and in taking in flank an attacking line advancing incautiously (*vide* R, S, T, in Map, p. 269).

Such a system was on many occasions adopted by us, and the Boers frequently had a few snipers carefully posted several hundred yards in front of their position, which made it extremely difficult for reconnoitrers to realize how the enemy was located. Night attacks could seldom succeed if such posts be out in front.

If good information be received of an impending attack, it may be very desirable, especially if at night, to place parties in ambush out in front, who may deliver a volley and retire.

Ambuscades. Ambuscades, *i.e.* occupying an unexpected position by a small party who suddenly open fire at close range on the advancing enemy, may often have important results—a small force so hidden away is able to inflict very severe damage with magazine rifles. The enemy advancing in the open is at a great disadvantage, and being taken suddenly by surprise, must become to a certain extent panic-stricken. Nor is this effect only

DEFENCE.

temporary, for once a force has been badly hit by an ambush, great caution will be taken in future, and dashing movements will not be carried out so keenly as before.

Talking of ambuscades, the following incident which happened in the Transvaal is not without interest. The handful of Boers that were in the habit of following our columns about the country, chiefly to look over our old camping-grounds to pick up lost ammunition and other useful odds and ends, had occasionally been ambushed by us. One day, soon after we had left a camp near a farmhouse, a small party of Boers was seen to arrive, and halting 300 or 400 yards from the house, started firing rapidly into it. This was evidently done in order to ascertain whether any party had been left secreted in the house, and no reply being returned, the Boers advanced and searched the ground about. Men forming an ambuscade should not be drawn into returning fire before the proper time.

The power of concentrated musketry and artillery fire has already been pointed out; but the importance of it should be well impressed on the mind. To *disseminate your target and concentrate your fire* must ever be a maxim. An evenly distributed fire on a line of troops advancing may cause casualties, but if not very heavy it is not likely to check the advance or bring about a retreat. A concentrated fire on one part of the line may bring about both locally, and hesitation soon spreads along a line.

Concentration of fire.

DEFENCE.

Referring again to the instance given on p. 83, it will be seen that the line occupied by the attackers is very scattered, and it can readily be realized how few bullets will strike them and how many be wasted in the intervals. If now the order can be given to concentrate the fire on one object, such as on the troops about D, it will be seen that this lot will suffer very severely and probably be beaten back, when the fire may be concentrated on another party of attackers.

Reservation of fire. Though long-range fire sent in the direction of an invisible enemy may do considerable execution, yet it will often happen that the most effective method of defensive fire tactics will be to reserve fire till the last moment, especially if not well supplied with ammunition. For the attacking troops will seldom offer much of a target till they get within 500 or 600 yards. This reservation of fire to the last was frequently practised by the Boers—as at Magersfontein, Modder River, Colenso, and other places. But even then they were usually too soon, the Boers being much afraid of a charge. The closer an assailant comes, the more effective should the fire be, and presuming the ground is perfectly open, and the defenders are not *very* widely extended, it should be almost impossible for the attack to succeed. If the defenders remain quiet and concealed till the enemy are within 100 yards, or even closer, the latter mask the fire of their own reserves and their artillery. If, then, the defenders suddenly open a very rapid fire, especially with

DEFENCE. 97

machine-guns, it should be most effective at so short a distance, and yet they could receive but little response. The enemy would probably be surprised and panic-stricken, their fire wild and uncertain.

The moment for opening fire would be just when the enemy were crossing a place where no natural cover existed. This was one of the great mistakes at Modder River. There, directly fire was opened, the advancing troops laid down among the scrub and were more or less out of sight. Had they advanced to a closer and more open spot it is difficult to believe that any could have escaped unhurt. At Magersfontein, the Highlanders got closer and suffered more. This method was insisted on with great success in the defence of Mafeking, where it was very necessary to husband the ammunition. At the start there were only 600 rounds per rifle, yet after many months of constant firing there were still 120 rounds per man at the end.

These tactics apply not only to musketry, but also to guns. The position of the latter should not be unmasked too soon. Boer "slimness" often misled us about their guns. Sometimes they never opened fire until the attack was well advanced, and sometimes again, stopped firing for long periods, leading us to believe that they had been "knocked out."

When exposed to a very heavy fire, especially of artillery, men will often crouch down under cover. But it is very necessary, under such Look-out necessary during attack.

G

circumstances, that a good look-out be kept, else, when the fire suddenly ceases and the men peer above the cover, they may find the enemy have approached close around them. On several occasions our mounted troops have rushed positions by taking advantage of this tendency. A British outpost was captured in this manner at Pienaar's River. At Talana Hill, too, it is said that the Boers crouched behind a stone wall that ran along the position, and thus allowed our troops to get close up. The Boers very often, under such circumstances, held their rifles above their heads on a level to fire over the parapet; a mode of action hardly to be commended, yet one that is likely to have more result than if no fire at all be discharged.

Supports and reserves. Supports and local reserves are even less necessary in defence than in attack. It stands to reason that it is best to have as many rifles as possible in the firing line. As has been pointed out, a gap of even several hundred yards is of but little moment now, so that casualties do not need replacing. A loss of 20 per cent. would be considered as very severe, and yet the absence of one rifle in five would not make a really serious weakness in the firing line. If at one place the garrison suffered very severely, it would not be advisable to push more men to that exact spot.

Such an act as reinforcing (in the ordinary sense of the word) a firing line under fire is most undesirable. With far-reaching rifles a range of a hundred yards more or less makes but little difference, and if the attack is being very strongly pressed on one point of the defence, it will usually be possible to bring up reserves to some point on the flank whence they can open a telling fire, without pushing them close up to the original firing line. For again it has to be remembered that a pressing attack only means that a mass of fire is being directed on one part of the line. It is no good reinforcing *that* with more men; the reinforcements would be sure to suffer heavily if they advanced up to it. The only proper mode of reinforcing is to bring a greater amount of fire to bear on the enemy.

It will be usual, with proper dispositions, to have ample warning of an impending attack, its general direction will be apparent, and the force can then be disposed according to the circumstances. It will, nevertheless, generally be desirable to have a body of troops in reserve to meet emergencies. These would usually be well back under cover. But though kept in hand, and ready for despatch to any desired point, such reserves might, if the ground were suitable, occasionally aid the general defence with long-range fire. Even these should be in extended order so as not to form a target for a chance shell. Positions for the reserves to occupy should be chosen beforehand. They would preferably be in rear of intervals in the front line. *Reserves.*

Troops in camp must, of course, be so disposed as to be readily got into position to repel attack. To stack away infantry in quarter column, though *Troops in camp.*

DEFENCE.

frequently practised even towards the end of the war, would seem against all principles of common sense. In the event of a sudden attack, not only do they form a target on which the enemy can concentrate his fire, but, crowded together, they are in the worst position for returning fire. To extend them takes much time, and would surely be attended with immense confusion in the case of a night attack. Infantry should certainly be camped in line facing outwards along the edge of the camp, and even with cavalry this should be carried out as far as possible.

Guns should, if possible, be in such a position that they could at once open fire. At Uitval the guns were placed where they were practically unusable, and the horses were shot down before they could be moved.

In small camps, and where an attack is at all likely, if the ground allows, all horses and animals should be kept in a hollow protected from fire; or, if such does not exist, they may be located on one side of a ridge with orders that should an attack come from that side, the horses should be at once taken over the crest to the other slope.

Standing to arms. Some generals kept up the practice of ordering all troops to stand to arms for an hour or so before daylight. This certainly greatly interferes with their repose, and must act detrimentally on their bodily efficiency through the day. If all troops are lying in position ready to repel attack, and if the outposts are alert, the force cannot be surprised;

but if they are packed together in among horses and waggons, without clear instructions, then, even though the attack may come when all are standing to arms, confusion and loss are sure to result.

Counter attack has hitherto been considered as a most desirable and even necessary adjunct to successful defence. But in order to study its applicability with modern weapons the exact meaning of the words must be more clearly defined. Formerly, the usual acceptance of the term implied a charge forward of the defending force to drive back the assailants with cold steel. Taken in this sense, it is hardly likely to be a typical feature of modern war. It would imply that the assailants had arrived within 200 or 300 yards of the firing line; at which distance musketry fire would be so deadly that the attack could hardly hope to succeed against a determined defence. If the defenders at this moment left their cover to advance, the tables would be turned, and a steady fire from the assailants should have a decisive effect, especially if well backed by their artillery. In the attack on Ladysmith on January 6th, according to the Official Despatches, "Colonel Hamilton, seeing plainly that the only way of clearing out those of the enemy's marksmen who were established on the eastern crest of Wagon Hill, within a few yards of our men, was by a sudden rush across the open, directed Major Campbell to tell off a company of the 2nd Bn. King's Royal Rifle Corps to make the attempt, which, however, failed, Lieutenant N. M. Tod, who commanded, being killed, and the men falling back to the cover of the rocks from behind which they had started." It is true that, shortly after, a counter advance by three companies of the Devonshire Regiment succeeded. "The Devons dashed forward and gained a position under cover within 50 yards of the enemy. Here a fire fight ensued, but the Devons were not to be denied, and, eventually, cheering as they pushed from point to point, they drove the enemy not only off the plateau but cleared every Boer out of the lower slopes and the dongas surrounding the position." But it will be noted this was, apparently, an advance from point to point, utilizing fire rather than the bayonet. A bayonet charge could only be advantageous if it were apparent that the attackers were absolutely wavering; when a dash forward might have a great moral effect, though even then a decidedly risky undertaking.

Such a counter attack with the bayonet could at best have but a temporary and local result. It would only be the front line of attackers that would be effected, and as it would always be usual to hold a number of troops in reserve, these acting temporarily on the defensive should easily be able to ward off the charge. Though no good instance can be quoted of a typical counter attack in the recent war (except possibly those on Wagon Hill), one can imagine what would have happened had such been attempted. For instance, at Magersfontein,

J. E. Middlemore]

BOER GUN-EMPLACEMENT ON TUGELA HEIGHTS.

[*Photo., Durban.*

after the repulse of the Highland Brigade, it might have been supposed that an opportunity occurred, but all eye-witnesses would probably agree that had the Boers issued from their trenches for a forward move, they would most certainly have suffered a severe defeat, coming under the fire of all the British troops in rear. If the defenders found their ammunition was running short, their only chance might be to charge forward as a last resource. But this course was not deemed advisable at Nicholson's Nek or other places.

The best form of counter attack would undoubtedly, in nine cases out of ten, be a steady, ceaseless musketry fire, which will be more deadly, more rapid, and, as a rule, more far-reaching than a charge with the bayonet.

In very broken country, or thick bush, where long-range fire and artillery are out of the question, hand-to-hand fighting may often be profitably undertaken.

At night, too, when accurate and distant shooting is impossible, shock action may be preferable. At Mafeking successful counter attacks were made after dark.

If an attack be repulsed, and the enemy has retired, an advance forward may be necessary to obtain a better field of fire; but this can hardly be considered a counter attack. The retreating, and perhaps disorganized, enemy may even be followed up, especially by cavalry, who might then have a good chance of inflicting a crushing defeat.

Mounted troops, watching their opportunity, may not only issue to attack by shock action, but they may often advance far away to the flanks of the attacking lines, and take up position from which they can bring a telling enfilade fire on to the attackers.

Retirements under fire. As a general rule to retire when under a heavy fire is a fatal move. It is a natural desire, when bullets are coming in thickly from the front, to run back to get away from them; but one cannot so easily get out of reach of long-range fire. If occupying a well-chosen position, it will nearly always be better to stand fast. Even if caught in a bad one, it may be preferable to lie still and maintain as heavy a fire as possible in return. By rising to retire the defenders are bound to expose themselves and lose heavily; their fire ceases, and the attackers are encouraged to persevere. If they stick doggedly to their posts, there is always a good chance, however numerous the enemy may be, that the latter will "cry off;" especially if there is any chance of the defenders being reinforced.

At Nicholson's Nek a company of the Gloucester Regiment occupied a sangar for many hours without much loss, but, finding the fire getting somewhat hot, a retirement was ordered. As soon as the men left cover they came under a terrific fire, and over one-third of their number were knocked over.

A commander, before ordering a general retirement, should take these points into consideration.

DEFENCE.

Even though occupation of the position may be deemed useless, it may be preferable to hold on till after dark in order that a retreat can be made without much loss.

There may be exceptions to this rule if there be a good covered retreat, and if there be a better position close in rear, although in the latter case it would have been better to retire to it before the enemy gets to within striking distance. If retreat in the open is necessary, it should be conducted by only a few men at a time.

Small parties, such as scouting patrols, or observation posts, may often have to gallop off under fire, but they will generally do so, if possible, before the fire becomes heavy.

In a general retreat or rearguard action, too, the case may be different, but in such circumstances the successive positions ought to be so chosen that, as far as possible, the retreat can be carried out safely under cover.

Since surprise has so great an effect, a very effective ruse may be arranged, especially if the position occupied is not a very favourable one for ordinary defence, by disposing a certain force somewhere in rear of the first line, and ordering the advanced troops, on a given signal to slip away, retiring as much under cover as possible, and clearing the front; so that when the would-be victorious assailants dash over the advanced trenches, they will be met by a withering fire from the second line.

Such a manœuvre was several times carried out by the Boers, notably in one of the Natal fights, where they abandoned their first position, and sadly surprised our troops with an outburst of fire from a second.

The great dispersion of the modern firing line renders it very difficult to control. If a long line is to be occupied temporarily, such as during a rear-guard action, and a retirement then made, it is by no means easy to convey the order so that the whole line will retire together. On such occasions a bugler sounding the "retire" would be likely to inform the enemy of just that information which it would be desirable to keep from him.

General retreat. To retire before a pursuing enemy, probably over more or less unfavourable ground, is a very different matter to acting on the defensive on a chosen position. Yet there are instances of a small force, on receiving information of the approach of a large body of the enemy, attempting a disastrous retreat, instead of, as would be a preferable course in nine cases out of ten, taking up a strong position, preparing it for defence, and holding on until relieved.

If short of provisions or ammunition, or if no chance of relief seems likely, no other course than retreat is perhaps open, but even then half rations and a careful expenditure of ammunition will last a long time, and a determined defence has more chance of success than a hurried retreat before superior numbers.

DEFENCE.

One of the most important considerations on taking up a position for defence is as to the means of retreat. This would be especially necessary if opposed to an enemy of greatly superior strength. If the retirement is necessarily limited to one route this will require to be carefully watched beforehand to guard against some of the enemy getting round to cut off the retreat. It may even be desirable to occupy commanding points beforehand with small parties to ensure the safety of the road.

The Boers were most successful in retreating before being beaten; but then they possessed the greatest mobility, and moreover decisive results could seldom come of such tactics.

A frequent practice with the Boers was, when finding themselves hard pressed, to disperse, and each man for himself gallop off in a different direction, or hide himself away, having arranged to reassemble at a certain place later on. Though such a manœuvre is not very suited to disciplined troops, and would not be easy to carry out successfully in a strange country, and might involve the abandonment of waggons and guns, yet there are circumstances under which such a course would be far preferable to an unconditional surrender or to the massacre of the whole force.

Scattering retreat.

CHAPTER IV.

THE SELECTION OF GROUND AND POSITIONS.

THE natural features of the ground and their suitability to various tactical operations is a subject which has hitherto, perhaps, not been sufficiently studied. In olden days of hand-to-hand encounters it mattered but little what the "terrain" was like, so long as it was more or less suitable for manœuvring over. If on a hillside the force nearest the top had an advantage over those below in being able to charge down with greater impetus than could be got by the others in clambering up; otherwise ground was of little consequence. As firearms increased in importance, and the distances at which opposing forces could harm one another widened, its influence in tactical operations became recognized. Now, when the deadliness of modern musketry and artillery fire is the ruling feature of war, when ranges have increased up to the limits of vision, and protection from the swarms of missiles most necessary, it is of supreme and ever-increasing importance.

Whether in attack or in defence, the selection

SELECTION OF GROUND AND POSITIONS.

of ground so as to enable the troops to bring the fullest effect of fire to bear on the enemy, to screen themselves from the hostile bullets, and to force the enemy to cross or occupy ground open to a sweeping fire, is the main object of the tactics of the battlefield.

In attack we must look out for folds of ground, valleys, beds of streams, ditches; even walls, hedges, and trees, under cover of which to bring the troops along, so as to approach unharmed or unseen within effective distance of the enemy; for suitable approaches along which supplies of ammunition can be conveyed to the advanced troops; for hillocks and ridges whence to bring a more or less commanding fire to bear on the position; for ground to be avoided as unfavourable to the advance such as that which is very open, boggy, or steep; for commanding positions for the guns, and good look-out spots with extensive view for the commander and for the signallers.

In the defence we have to select positions round which the ground must be unfavourable for the advance; for cover behind which the riflemen may crouch, and horses and waggons be secure from fire, for covered communication with the rear, and, as before, suitable positions for the artillery and look-out posts.

Taking first the larger natural features of the ground, we have *shape of hills* and undulations. <small>Shape of hills.</small>

The general sectional shape of a hill varies greatly, and on this depends its suitability for occupation.

110 THE SELECTION OF GROUND

The slopes will usually be more or less either straight, convex, or concave, as in the three instances depicted.

In the first instance, fire from the top will sweep the hillside. In the second, in no position can all the side be clearly seen, and much "dead" ground

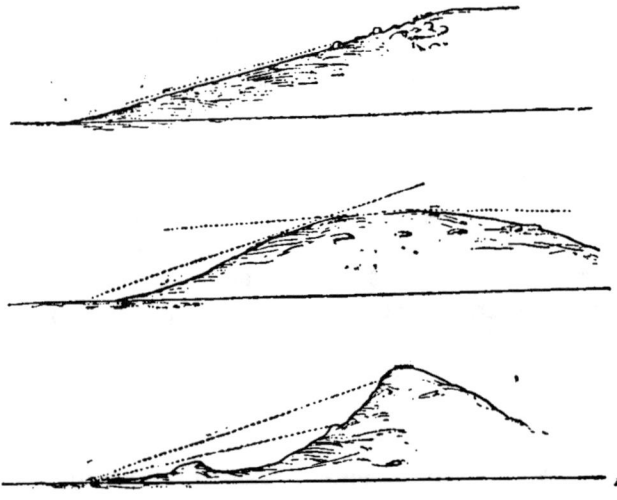

PROFILES OF HILLS.

will exist. In the third the fire will be "plunging," but the ground is very suitable for several tiers of fire, which is impossible with the other two if firing on an enemy advancing up the hillside.

The second illlustration represents a large, isolated, rounded hill at Kaalfontein (near Krugersdorp). This was chosen for a fortified post. Guns

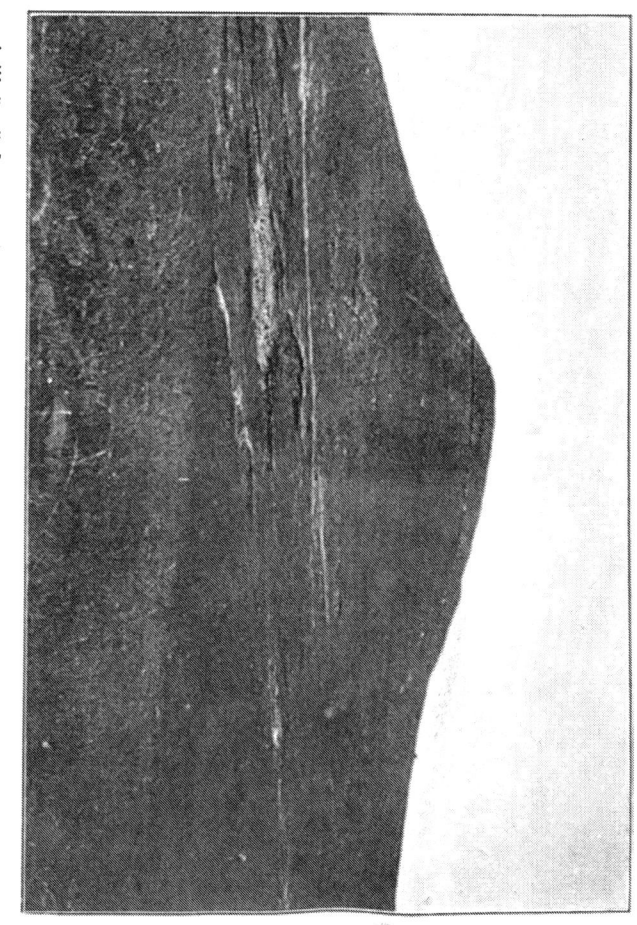

J. W. Bradley] CHILD'S HILL, SHOWING CONVEX AND CONCAVE SLOPES. [*Photo., Durban.*

were placed on the top, so that they could sweep all round. But from this point nothing could be seen of the plain below within about 1000 yards of the base, so that once an enemy approached within this distance the guns became useless. Trenches were constructed around the hillside some distance down. These were quite out of sight from the top of the hill (so that a commanding officer could not control the defence or watch the progress of an attack from there). Yet further down was a large extent of " dead " ground invisible from the trenches. Moreover, there was a very small field of fire immediately in front of them. Had the trenches been placed at the bottom, the line of defence would have had to have been about doubled in length. This, then, might be considered a bad shape of hill to choose. Yet, on the other hand, it possessed certain great advantages. Thus the enemy, on coming within 1000 yards, could not fire with aimed musketry on the guns. Secondly, if parties of the enemy got through the line of trenches, the occupants could turn and fire towards their rear without fear of hitting the gunners and men of their own side on the top. The enemy ascending the hill would very suddenly come under the close fire from the trenches.

Near Leeuwberg, south of Bloemfontein, was a rocky hill of the concave shape, shown in Fig. 3. This was occupied by outposts, and on the top was erected a small fort for headquarters, signalling station, and supply store, while at various advantageous points on the side, schanzes and breastworks were put up; all of which were fully visible from the top, and commanded all the ground below.

Often two of these forms will be combined. Thus, at Talana Hill, the lower portion was comparatively smooth and glacis-like, the upper portion steep and broken. The Boers occupied a position near the top. The result was, that though our loss was heavy in crossing the open, once the bottom of the steep part was gained, cover could be got, and the forces were on an almost equal footing. Had the Boers been posted at the head of the slope, their grazing fire would probably have proved too deadly to advance against.

Magersfontein presented a somewhat similar section, but there the trenches were placed at the base of the rough kopjes—with telling result.

Sometimes the curve of a hill may happen to correspond more or less with the trajectory of the rifles; so that bullets passing over it would graze along the ground and prove destructive to a force advancing unseen up the reverse slope.

This occurred at Belfast, where we had serious losses from the unseen foe over the hill. But this, of course, must be rather a chance characteristic and one very difficult to measure.

If the side of a hill be "straight," its suitability for defence will entirely depend upon the surface. If it be smooth grass, the fire from the top will be grazing and very effective, but if, as often happened in South Africa, it be rough and rock strewn (not easy to discern from afar), plenty of cover is available to an assailant.

But the dimensions of the hill will greatly affect the suitability or non-suitability of such characteristics. A very large rounded hill implies large areas of dead ground, whereas quite a small hill of this shape may be very favourable.

Rounded hills are generally suitable for artillery, but not for infantry. If the surface be broken a concave slope should be chosen.

The possession of command of ground, which may be defined as the occupation of a position of higher elevation than that occupied by, or likely to be occupied by, the enemy, is generally looked upon as not only desirable, but of the greatest importance. Yet this is a subject seldom carefully considered, and the exact advantages not taken into account. *Command of ground.*

The actual benefits to be gained by the occupation of commanding ground are—

1. A good view may be obtained to watch the movements of the enemy.
2. Riflemen may be able to see and fire at men lying flat on the ground or behind low cover, such as unevenness of the ground.
3. Gunners can better direct artillery fire and note its effect.
4. As a rule, all in rear of the firing line is under cover from the front, so that retreats,

reinforcements, and carriage of supplies can be carried out unseen.

Other minor advantages are that an enemy is more or less delayed clambering up a steep slope. It may be possible to arrange several tiers of fire to defend the hill, and a certain moral effect of superiority and safety is felt by those in the more elevated position.

But if this be a correct summary of the advantages, it can soon be shown that they are not of great importance.

In considering what is gained by the first characteristic, it must be remembered that there is no necessity whatever for the privates or even subordinate officers to observe the general movements of the advancing foe. It will be quite sufficient for them to watch over their immediate field of fire, and so long as they can see about 800 to 1000 yards to their front, they need nothing more. Objects at this distance can be seen on a flat plain, without being elevated above it. It is only to the look-out man, to the general in command, and to the artillery officer directing the fire that a distant view is desirable. For them a small steep hill, a church tower, or a balloon may suffice. If the country be wooded or much broken up, an elevated observation post is of comparatively little value, since an enemy can then approach unseen.

The second consideration is very dependent on circumstances, and much will depend on the exact nature of the surface. If the ground below be strewn with large boulders and rocks, such as frequently occurs in South Africa, good cover may still be obtained despite the inferior elevation. Ditches, dongas, walls, even bushes, may practically nullify the advantages of command. It is a matter of degree, depending on the steepness of the hill, the range, and the surface of the ground.

At Koster's river (where a number of Bushmen were hard pressed and suffered considerably) the main Boer position was on some low rocky, wooded kopjes. We got on some high open hills on their flank which completely "commanded" their position in the ordinary acceptance of the term. Yet one could see nothing whatever of the enemy carefully hidden away among the rocks and kloofs. The reports of their rifles were the only indications of their presence, and after a time they all got away practically unseen.

As regards cover for those attacking commanding ground (or defending the lower ground), one is apt to forget what a very small angle of elevation most hill features present at a distance. Thus, if a hill be 300 feet high, and a rifleman on top fires at a point on the plain 600 yards off, the bullet (without considering trajectory) would descend at an angle of only 1 in 6—that is to say, that, if passing just over a bank 1 foot high, it would not strike the ground till 6 feet in rear of it. Therefore, unless the command be very considerable as compared to the range, ordinary irregularities of ground offer good cover, and it may be said that *intrenchments cannot be commanded* if properly defiladed (which, as a rule, would seldom necessitate an increase of more than a foot to the height of parapet).

Again, the ground to the front greatly affects the efficiency of command. If the ground slopes gently away from the position for several hundred yards this does away with the superiority supposed to be gained by elevation; for any little depression or mound on this gives cover just the same as if it was a flat plain. Therefore, as regards musketry, to have full advantage of command, the ground to be crossed by the enemy must be flat and smooth.

In comparing the efficiency of a position on a commanding site with that of one on the flat, we must consider the effect of the fire.

The fact of having a more or less bird's-eye view of the ground below implies that the fire must be plunging, which is objectionable.

In firing from a height at an individual below, if the bullet goes over his head or to one side or other, or even short of him, it goes plump into the ground. Now, if that bullet be fired from on a level, if it passes over the top of the target or just to one side, it may travel on for hundreds of yards, and eventually it may chance to find during its course, a billet in some one behind. Even if it strikes short the chances are, fired so flat, that it will ricochet on to the mark. When a general hail of fire is the object this becomes a matter of importance.

There are, too, even advantages in shooting *up* hill. The opponent is then likely to stand out against the sky-line, and so be visible even at long ranges, infinitely more so than one downhill standing against the ordinary background, to which it is now the fashion to similarize the uniform.

As for the third advantage of command, *i.e.* for artillery, it must be owned that a position with a good view is, without doubt, a desirable quality, since not only is it necessary to see the enemy's position and movements at a considerable distance, but the effect of the shells should also be noticeable, especially as to whether they pass over the target or fall short of it—often a difficult matter to judge when seen on a level. The greater the command the more discernible the effects would be.

And yet even here command is not a *necessity*, for though it may be desirable for each gunner to see the exact effect of his shot, the fire *can* be directed by one observer, and it is not essential for the actual guns to be placed on a hilltop.

As regards the object of occupying the top of a hill to allow communication, reinforcement, retreat, etc., on the far side, it is again evident that elevation is no great advantage. If the hill is liable to be surrounded, or even out-flanked, this will be negatived. Moreover, a donga, or deep valley in a plain, would form as good, if not better, means of communication. Such existed at Modder River and at Sanna's Post. In fact, the cover in rear depends on the ground there falling away, rather than the position itself being high. The mere fact of elevation implies nothing: a ridge or bank only eight or ten feet high in a plain would cover all behind it.

There is one point in which a position on commanding ground has an advantage over one on low-lying ground with a depression behind it, and that is that a high ridge may cover *all* the country behind it for many miles, whereas the ground in rear of the other may be visible. This, however, is not usually of great importance.

As for high ground enabling the defenders to employ several tiers of fire, it must be remembered that if the elevation is great it also enables the rear line of attackers to fire over the heads of the front lines at them, and this is an important point in attack, as, if a very heavy musketry fire can be kept continually pouring on the position, the attacking line can advance with comparative ease.

At Modder River the rear lines of our troops considered it very risky firing so close over the heads of the front firing line that it was seldom practised during the day, and the defenders were able to maintain such a fire as to prevent any advance.

To obtain the advantages enumerated to any appreciable extent, the position must be very considerably higher than the surrounding ground. The fact of one hill being very slightly higher than another confers but little benefit on the force holding it. Yet one often hears of a hill being "quite untenable" because it is "commanded" by one slightly higher. Thus, if a force occupy a position at A, it will in reality have no great advantage over one at C, while one at B will be actually worse off since it has no cover in rear—actual elevation is not material. The side of an open hill will be exposed to those in the plain below, just as much as those below will be to any one on the hillside. If the diagram be held so that the line DE is horizontal this may be better realized, for then C will appear to command A:

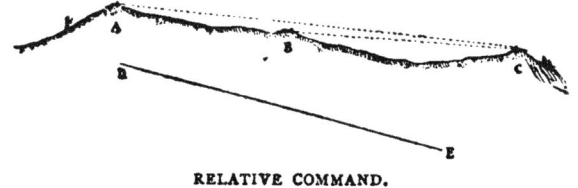

RELATIVE COMMAND.

Our position at Spion Kop was somewhat like this, the Boers occupying a lower feature of the hill, as at C, and were no worse off than our troops on the top.

At Roossenekal the village is, in the ordinary acceptance of the term, commanded by a big, gently sloping hill, of which the profile would be somewhat as in the cut overleaf. To defend the town with a small force the idea most natural would be to occupy the top of the hill, A. But there were several objections to this. First, it was far from the water supply, C. Even if a post were formed at the bottom to protect the water, it would be difficult to convey the supplies up the open hill under fire from the plain. Secondly, the far side of the hill was much broken and uneven, so that an assailant could easily get to close quarters. It would then be preferable to occupy a position by the houses at B, whence the face of the hill would be swept with fire and good cover obtained among the buildings.

The Boer position at Modder River was situated on what may be called the ridge or crest-line formed by a deep river-bed with a flat open plain

POSITION AT BASE OF SLOPE.

extending far to the front. This formed a very strong position having a wide field of fire, good covered communications, cover for horses, waggons, etc., plentiful water supply, soil suitable for trench digging. Yet it lacked the great object of command—ability to see men lying behind low cover. Had a small hill or any good look-out place existed, the position would have been well-nigh an ideal one.

If, then, the country in front be flat and open, a very slight elevation is all that is necessary. If covered with low bush or rocks, a height of a few feet more is desirable. If the ground be broken or undulating, greater elevation above it must be looked for, while if very cut up and covered with trees, no amount of command will avail to give the desired view. So that a really high hill, especially for musketry purposes, is quite unnecessary. Cavalry and mounted troops will be impeded in their action by steep slopes, and it will probably involve much difficulty and labour to get artillery up them; so that, unless the position on the top were exceptionally desirable, even artillery would not be likely to utilize so elevated a height. Its lower spurs may be required for occupation, and if so, a small force may be necessary to prevent the enemy gaining the summit.

It seems, then, that the idea hitherto held of an ideal position being one on a high hill or ridge, is fallacious. This is a very important point, and it will be desirable to carefully investigate all instances in the war in which a large force was attacked when in position either on a hill or on a flat plain. The following list gives the general result:—

	Position.	Result.
Talana Hill	High and steep hill	Position taken.
Elandslaagte	Hills	,, ,,
Nicholson's Nek	High hill	,, ,,
Belmont	Steep hills	,, ,,
Graspan	Large hill	,, ,,
Pieters	High hill	Summit gained.
Willow Grange	Steep hill	Position taken.
Vaalkranz	Hill	,, ,,
Driefontein	,,	,, ,,
Stormberg	Big hill	Attack repulsed.

	Position.	Result.
Modder River	Flat plain	Attack repulsed (for day).
Magersfontein	Trenches at foot of hill and on flat, artillery on hills	Attack repulsed.
Colenso	Trenches on flat, artillery on hills	,, ,,
Paardeberg (Feb. 13)	Flat plain	,, ,,
Mafeking, assaults on (as well as counter attack on Game Tree)	,,	,, ,,
Sanna's Post	,,	,, ,,

(Such positions as those on Spion Kop, Wagon Hill, and many others, cannot be distinctly classified as hilly or flat, the chief fighting having taken place on top of the hill.)

Though various circumstances may have contributed to the issues of these fights, the fact remains that, out of ten attacks on commanding positions, nine succeeded, and only one (Stormberg) was repulsed; whereas out of six attacks over the flat none succeeded, which at all events proves that commanding ground is not a *sine quâ non* to successful defence.

A ridge of hills some little distance *behind* the main infantry position will, however, be of value. The guns may be on the slopes, observation posts can be established on the top, and camping-grounds and covered lines of retreat arranged in rear of them.

So that it may certainly be said that the best position is one where a flat plain extends to the front, while hills rise immediately behind.

In the very minor operations command of ground may be important. A small force moving under a steep hill is, to a certain extent, at the mercy of a few men firing down upon it; but where it is a matter of a line of defence being attacked at varying distances up to a thousand yards, a hill is of but little advantage.

Hilly country not only causes extra exertion on the part of the attackers to surmount, and gives better opportunities for movements in rear of defensive lines, but it also bestows another great advantage on the defence, that it makes reconnoitring more difficult. In some parts of South Africa where kopjes were numerous, it became hopelessly tedious sending scouts clambering up kopje after kopje, and seriously delayed the main advance. Hilly country.

It is to these causes probably that the difficulties of the Natal operations may be ascribed.

But for other purposes hilly country is not so suitable for defensive operations as a flat plain. In broken country it is very seldom that a really good field of fire can be obtained (and this is, nowadays, all important), and the attacking force is enabled to approach by taking advantage of the valleys and depressions. On hilltops water supply is usually a difficulty.

Though the larger natural features of country may, on first thoughts, appear to have much importance as regards their adaptability to tactical manœuvres, yet there can be no doubt that the movements of troops and the effect of their fire Details of surface.

are much more dependent upon the detailed nature of the surface.

Such characteristics may be summed up as follows:—

Flat and open country gives every advantage to the *defence*. The defenders can see any one approaching from afar, and can sweep the ground with their fire. Even if no cover is available, troops lying down are practically invisible at a distance, so that they would not be likely to suffer nearly so greatly as would an advancing force.

Undulating but open country, while enabling positions with good fields of fire to be taken up, yet may often enable the attackers to approach by the valleys.

Rocky soil renders the digging of trenches difficult, but if the ground be strewn with loose rocks, good cover can be rapidly built up, and head-cover contrived. Yet detached rocks or boulders in front provide cover for the attackers as well. Bullets ricochet off the rocks at odd angles, and splinters of rocks are thrown around both by bullets and especially by high-explosive shells. This tells more against the defence than against the attack.

Long grass, crops, and scrub favour attack, as advancing troops are not so conspicuous, and are invisible while lying down to rest. They can often creep up to a position practically unseen. The Boers approached close to our intrenchments at Warmbaths through the long grass, and after a considerable amount of firing, were able to creep away unseen. At Modder River a sparse crop of thistles and karoo-bush sufficed to hide the troops lying down. It is worth remembering that in a short space of time the same ground may rapidly change its nature by fresh crops growing up, long grass being burnt away, or artificial clearings being made.

But detached bushes and clumps may, if under the clear view of the position, even be favourable to the defence, since it gives a false sense of security to troops attacking, who may crowd behind such cover, although not really protected from the fire.

Sand, plough, gravel, etc., form surfaces in which it is easy to scrape some cover, even without special tools. This would favour attack as much as defence.

Boggy and water-covered ground is often looked upon as impassable; but this should be carefully investigated. There were several instances of attacking Boers wading through water up to their waists. I was once assured by an officer, who should have known better, that there was no danger of attack from a certain quarter, as the ground there was all boggy; but on going over it, I found the mud nowhere more than ankle-deep.

Thick bush and woods render advance difficult, extended troops being apt to lose touch, and the defensive position difficult to locate; also cause progress to be very slow, owing to the difficulty of reconnoitring. In the "Bushveld" on the northern line to Warmbaths, it was found very difficult to work mounted troops rapidly.

Bush which is thick at a height of two or three feet and above, but has nothing but bare stems below, enables defenders lying down to watch and fire at the legs of advancing foes, who are unable to see to fire back.

Woods, clumps of trees, and even many detached trees spread about, give to a position facilities for hidden means of communication. The position occupied by the Boers on the banks of the river at Paardeberg was covered with trees, which rendered it extremely difficult for us to make out their dispositions or note the effect of our shell fire. Bush, then, generally favours defence.

Defence of woods. Woods have not perhaps so much tactical importance as formerly. If the edge of a well-defined wood be occupied, it forms a mark for the enemy's fire, and the absence of a target is one of the greatest difficulties in the attack. But the edges of a wood are not always well-defined. If it thins out gradually it may better conceal the exact position.

Though some cover from fire may be derived from large trees, it is deceptive and unreliable. Our troops suffered greatly through massing in the plantation at the foot of Talana Hill. At the engagement near Maritzani, on the road to Mafeking, which was fought in a wood, the mere sound of the bullets smashing down branches and ripping bark did not add to the peace of mind so desirable in defence.

There is one more point in which operations may be said to be affected by the country, and that is the extent of the knowledge of it possessed by the troops. When every man is perfectly acquainted with the district a very great advantage is gained, but this advantage will, of course, vary with the nature of the country. Where it is perfectly open and with flat plains or rolling downs, the advantage of being acquainted with it is of comparatively little importance. But when warfare has to be conducted in broken and woody country, especially where large and precipitous hills exist, familiarity with them is most invaluable, and may have a great influence on the results of operations. Many of our difficulties in South Africa can be ascribed to lack of this knowledge. *Knowledge of the country.*

Having thus considered the general natural features of country, we may now investigate their adaptability to a particular object, such as a defensible position. *Selection of position.*

In selecting a locality for defensive operations there are two different points to be considered. There is the position to be occupied by the defenders, and there is the ground to be crossed by the attacking force. The latter is often the more important consideration. There are many instances of troops occupying a so-called good position, suffering severely because the enemy was able to creep up close under cover, or was able to bring an overwhelming fire to bear on the defenders.

If perfectly open ground surrounds the position it is impossible for an enemy to approach (by day or even by night) without incurring very heavy loss.

In selecting any position for more or less permanent occupation, the following, in their order, should be looked for :—

1. *Water supply.* This is most important, since, if a force has none or becomes cut off from it, it cannot hope to hold out twenty-four hours, no matter how good the position in other respects. Several instances have occurred where this has been the case. If no natural supply exists, wells or ponds may sometimes be made. (At Rustenburg, where a small stream ran through the town, a large pond was dug out and dammed up so as to form a good supply.) Otherwise tanks, barrels, tins, or other receptacles must be provided and filled. It is essential that these be screened from fire. When the Boers occupied the fort at Mafeking, the only water supply was from an iron tank. Knowing this, our men fired perpetually at the tank and riddled it with bullets, letting the water run out, and this alone would have forced the Boers to surrender. At Magersfontein water supply must have been a great difficulty, but large tanks were erected at intervals behind the kopjes, so that at night the men lining the trenches could go out to fill their bottles. Could we have occupied positions

on their flanks this would probably have necessitated the evacuation of the place. At Modder River and at Paardeberg the Boers selected positions on the river, doubtless chiefly to ensure water supply.

2. *Good field of fire for musketry.* The ground to the front and flanks (and usually to the rear as well) must be *flat*, that is, without undulations and depressions, and its surface must be *smooth*, that is, without rocks, banks, thick bush, or high vegetation, and this open space must extend for at least 200 yards from the position, and as much further as possible. The nearer the line of defence is approached the more important does this absence of shelter become. A position having a clear space of 200 yards in all directions is, as a rule, much preferable to one with a good field of fire for 800 or 1000 yards on one side, but broken and daed ground closely approaching it on the other; but if the country beyond this range be broken and covered, and there be no good cover available for the defence, the enemy might lie concealed in this and keep up a hot fire, and so render the position untenable. Yet such open ground is not always, in practice, so easy to find as one may imagine. A few scattered trees may help to conceal the position.

A river or sheet of water forms an ideal frontage, affording absolutely no cover, and practically impossible to cross rapidly without exposure and noise.

3. *Natural cover* for the firing line, or suitability of soil, etc., to making artificial cover. Banks, mounds, and walls unsuitable for utilizing as parapets, may often be taken advantage of for protecting from a frontal fire troops who, in turn, are able to bring a *flanking* fire along the front of the position. Sometimes positions for reserves or special bodies may be taken up on the reverse slope of a hill, whence only the tops of some trees or other objects to the front can be seen. If the enemy is then reported to be approaching these trees, they may be fired on with rifles sighted below the proper range, and the bullets will fall at the base of the trees, the firers meanwhile remaining quite invisible.

4. *Cover for reserves*, communication, and for horses and waggons. This would usually be mostly natural cover, such as depressions in the ground, or even belts of trees. But if such does not exist, and *if the soil is favourable*, trenches and pits can be dug, or big walls built up, for their protection.

5. *Good view* for commander, look-out men, and for officers directing the artillery. It is very necessary to be able to watch the approach of the enemy from a distance, and so to be able to anticipate his attack. It will often, too, be desirable to communicate by signal with other forces at a distance. A church tower, a housetop, high tree, or even a specially built tower may afford a good enough look-out place; but these, especially the last-named, sometimes form conspicuous targets for

the enemy's fire, and are then unsuitable for their purpose.

At Mafeking a wooden erection was built up on top of the headquarter house as an observation post, and others were erected at other points. In a big position there should be a number of such look-out places. If a balloon is available, the perfection of a look-out post is obtained.

6. *Field of fire for artillery.* This, though desirable, is not really very important if there be a good observation post, whence orders and information can be readily transmitted to the gunners Shells can be thrown into hollows or into the reverse slopes of hills if an observer can direct their fire.

Besides the above, it is desirable to have—

7. *Obstacles to advance of enemy*, such as a river, bog, or even scrub, etc., which would delay him under fire.

8. *No good position for enemy's guns.* If he be comparatively strong in artillery a good position, especially for a small post, may often be found on the centre of a plateau half a mile to a mile across, or if it be surrounded by thick woods 800 yards or so from the position, the guns could hardly come into action so close to the defending infantry.

At Ladybrand the British position was on top of a high hill, so that the Boers in shelling it were unable to see from below the effect of their fire, else greater damage might have been done.

9. *Absence of observation posts* for the enemy, whence he can direct his fire and note weak spots

in the defence, may also be an important consideration.

10. *A good line of retreat*, covered as far as possible from fire and especially from view, should always be chosen. It may often, for various reasons, be necessary to withdraw a portion of the forces, even though in good position. It will be all the better, as disconcerting to the enemy, if this line of retreat be not directly to the rear. If the force retires and takes up positions now to the right rear, then crossing over to the left rear, and so on, the enemy will be much puzzled as to the exact line of retreat.

11. *Supply of fuel* has often proved a difficult problem. If no woods exist, a large force will very soon use up any small supplies, and it may be impracticable to import much wood.

12. *Grazing* for horses and cattle is another important consideration when a force has to remain in one place for some time. This should, for obvious reasons, be within the line of outposts, and if possible, protected from the enemy's fire.

Selection of temporary position. If a position has to be taken up rapidly, as during the course of an action, and for temporary purposes only, the order of importance of characteristics will be different.

It should be well impressed on the memory that the following is the order of importance:—

1. Clear field of fire for at least 200 or 300 yards all round.
2. Cover for firing line.

3. Cover for horses, waggons, etc.
4. Line of retreat (especially in rear-guard actions). Then look out for the artillery positions, water supply, etc.

By "a position" will usually, nowadays, be implied a series of small independent positions, mutually supporting one another. Care must be taken in so selecting and occupying them that each is able to hold out independently of others. On several occasions we have heard of one post being the "key of the position." At Dewetsdorp the abandonment of one small post rendered the whole position practically untenable. The same occurred in several other affairs. With a little care separate posts can nearly always be arranged to be unaffected by other works falling. *Occupation of position.*

A position is often taken up with the idea of facing one way only, with front, flanks, and rear. It is much better to arrange that it is equally defensible from *all* sides; so that, if attacked in flank or rear, it will not necessarily be taken at a disadvantage. This should, of course, always be the case with small independent posts.

It is sometimes argued that a work cannot be attacked in rear because some other work is there, but, as has just been pointed out, it is always better not to rely on the support of other works. Moreover, though not actually attacked, artillery or long-range musketry fire may be employed to take the work in reverse.

A position on a plateau is not only advantageous

to avoid the enemy's guns, but the edge or ridge makes a good position for the defenders' outposts, who can then greatly delay the attack; even guns may take up a first position there, and then retire. It should be practically impossible to reconnoitre such a position. If the edge of the plateau be not extensive, the numbers of the attacking force will be limited. Thus, if it be 1000 yards round, not more than 1000 men could be utilized in the attack at one time.

If the position is not an "all-round" one, the flanks should be very strong. A small and quite detached position may often be taken up, if naturally strong, well away to either flank.

It is undoubtedly most risky to occupy an unknown position after dark, or even during fog. Features of ground appear so very different in the dark that, when morning breaks, one is often quite surprised at the unexpected outlook. This has often to be done in taking up outposts, but it certainly should be avoided if possible. In fog, too, when nothing is to be seen further than a 100 yards or so, it is very risky. Spion Kop taught us this lesson with severity. From a distance the position looked favourable, so it was taken in the dark and fortified during the morning mist.

Though it will usually be best to arrange that no one work is the "key of the position," there are occasions when it may be desirable that some central position of a group may dominate the others. Thus, if a small hill is occupied, and a gun placed on top with a reserve, while other men are scattered in a chain of trenches lower down the hill, it becomes a question whether these trenches should be enclosed works or not. If one of them falls into the hands of the enemy, it might be necessary to be able to drive him out again; the centre work is very unlikely to be captured while the others hold out. On the other hand, if each trench be protected from all sides, it would be very difficult to capture all of them, and then, even though the "key" were taken, one or two trenches holding out might make it practically untenable.

In choosing a position, the particular features will vary in importance, according to the object with which it is to be held. Sometimes it will be necessary to occupy one liable to be attacked from any side; sometimes from one side only. In some cases it will be necessary to enclose a certain area of ground; often the choice of ground will be greatly restricted. Horses, waggons, and stores may have to be protected, though this is not always required. *[margin: Positions vary with object.]*

The following are the principal objects with which a position is held, together with the usual circumstances and dispositions:—

"*A*" *for long occupation.* (Can generally be greatly improved artificially.)

1. *To defend a town.* Chain of positions selected around. Usually can rely on town for supplies. Waggons and horses mostly kept in town. Main

attack only towards one side of each position, *i.e.* away from town and other works (though to be independent, dispositions must always be made in case of being *temporarily* surrounded). *Must* occupy positions around, however unsuitable they may be naturally. Cannot unduly extend line of defence, yet must be well in front of town.

2. *To defend depôt.* Often supply stores and rest-camps are required at various points on line of communication. Usually considerable latitude allowed in selecting a good position, somewhere near main road. Must be "all-round" position, good water supply, plenty of "protected" space for stores, convoys, and troops passing through. For the latter reason undesirable to have one small post on a plain. Near Middleberg a column came to such a post. They camped all round the fort, thereby entirely masking fire from it and rendering the work absolutely useless for their protection. At Holnek the position consisted of several small hills, completely enclosing a valley with stream running through it, which was eminently suitable for camping any passing troops. Each hill was strongly fortified.

3. *To defend post*, such as a railway station, bridge, etc. Choice of position strictly limited, however bad it may be from a tactical point. Though sometimes a good position close by may so command the country round as to be practically sufficient.

Here an all round position has to be occupied. But not as a rule necessary to provide protected space for convoys or stores, or even for any waggons and horses.

4. *To defend road or pass.* More latitude allowed than last. Usually a very commanding position with good views along road desirable. "All-round" position. Water supply often a great difficulty in hilly pass and commanding position.

5. *Strategical position.* Selected with care solely to combine, as far as possible, all that is desirable in an ideal tactical position. Usually in form of long line of posts, providing for main attack from one side only.

"*B*" *for temporary occupation.*

6. *To protect camp* of troops on the march, will usually be limited to within a mile or two, as depends on length of day's march, etc.

Good camping ground is usually the first requirement, but fighting capabilities are too often sacrificed for this. Generally, it is sufficient to select a good defensive position for the chain of outposts, while the main force is sheltered from rifle, and, if possible, also artillery, fire. Main attack on the posts chiefly from outside; but here also important to be prepared for attack in rear.

A good position may often be found at the head of a valley. The outposts on the ridges may be fairly close in, and only one side of the camp exposed to distant fire.

A good water supply—preferably close at hand—for men and animals is usually a necessity. This, unfortunately, is seldom to be found on or near a hilltop.

Grazing for animals and fuel for cooking is often also a very necessary consideration.

If a strong night attack is apprehended by a small force, it is advisable to shift its position after dark to a previously selected one, leaving fires burning.

However strong a natural position for defence by day, it will almost always be necessary to have a ring of posts in close connection to be sure against a surprise by night. But in many positions, such as a flat plain, or one commanded by hills near, this is not enough, as protection from snipers and rifle fire at long range is necessary; which protection can only be got by pushing out the chain of posts much further from camp, or by having posts well out in front of the chain. But if the position be on the *top of the hill*, especially if it be flat-topped, or in the centre of a *ring of small hills*, which can be occupied, the camp will be practically safe from aimed fire, and a chain of posts around the hill or hills will keep all quite secure. If a complete circle of hills cannot be got, even a semicircle will be good, or a simple ridge may give good cover towards *one* side. There should be no commanding hills within rifle range. The crest line of the hills or *enceinte* of hills should be of as small a perimeter as possible considering the area of ground necessary for the camp. It is usually unlikely, if for one night only, that guns will be used; otherwise desirable to have good positions for guns, and no commanding positions near for enemy's guns.

7. *Defensive tactical position* during an engagement. Usually not much time or opportunity for careful selection or of artificially strengthening. But these should be attended to as far as possible.

One of the finest positions occupied in South Africa was that at Magersfontein. Here were to be found practically everything that can be wanted. A line of kopjes, mostly low and steep, but rising to a considerable elevation at one end, ran roughly north and south for about four miles, among them were made good gun-emplacements. The ground in front was open and gently sloping away for many hundred yards. *A typical position.*

A large well with windmill pump existed at the north end of the line, and the position stretched away down to the Modder River. Some large iron tanks had been procured, and were placed at intervals behind the line of kopjes. The glacis-like slope to the front afforded a perfect field of fire for the riflemen. A fine open view of the country was obtained from the trenches as well as from the top of the ridge, along which emplacements for artillery were constructed of the boulders. The soil at the foot of the kopjes was most suitable for digging, and deep under-dug trenches were there excavated. There was an excellent field of fire for the guns, and good cover for reserves, camp, hospitals, etc., just behind the kopjes, several roads running away from them to the rear. The country to the front was very open without any commanding hills, yet there was a sufficiency of small trees and scrub to provide fuel, and plenty of covered grazing ground in rear.

Considering all points, it may be laid down, almost as a rule, that in occupying a position, infantry should never be posted on a hilltop (except as reserve or guard over guns or stores, or, of course, as a small independent post for any purpose).

Guns and look-out posts should be on the hilltops with the infantry ranged round the base or lower features of the hills.

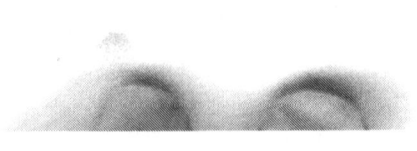

CHAPTER V.

FORTIFICATION.

SINCE modern fighting consists almost solely in pouring in the greatest possible number of missiles with the object of rendering inefficient the greatest possible number of the enemy, it becomes a matter of the greatest importance to protect one's men from harm, and when no natural cover exists, it is imperative that some should be made.

The Boer war has not been one in which the value of intrenched positions has been thoroughly shown. Though many instances abound, sufficient to prove the great advantage of them, yet they have not had so crucial an importance as it is very probable will be the case in future wars. The Boers, who often intrenched their positions, generally abandoned them as soon as their flanks were threatened. We also frequently made strong field fortifications, but these were comparatively very seldom attacked with energy, so that their value was not made apparent. The attacks on Magersfontein, Itala, Colenso, Paardeberg, Mafeking, and

Wepener demonstrated the great value of artificial cover.

Time devoted not wasted. During the early stages of an attack the defenders may often find that devoting time to intrenching may be preferable to utilizing it for firing back at the advancing enemy at long range. The damage to be done to an enemy is not likely to be serious while he is still 1000 yards or more away, and when he has come to close quarters, every bit of cover will be invaluable to the defenders. If a small force find itself being overtaken by a more powerful one, it would be better, while engaging the enemy's attention with a few men, to select and well intrench some good position, and there await his attack, rather than to spend the time operating against him and endeavouring, probably without any success, to drive him off with fire alone.

Complaint is often heard about the unnecessary labour of intrenching when no enemy is apparent, but the importance of it is very manifest when the bullets begin to fly, and men's lives are dependent on the cover available.

On January 6, two months after Ladysmith had become invested, Wagon Hill still remained practically without artificial defence. The result was, that though the Boer attacks on that day were elsewhere repulsed with ease, here desperate fighting took place.

Old and new systems. One of the most notable features of the war has been the change which field fortification has undergone. The teachings of the older text-books have been cast aside and practical experience has developed quite a new system.

It was once said that the best authorities on field fortification were not the Engineers, but rather those who were most frequently under fire and well versed in the knack of taking cover. This is more true to-day than ever, and the infantry-man, who has evolved the present system, must be the most practical authority on field intrenchments.

Once fortification was an art—architecture. It then became a science—military engineering ; but now it has become but a branch of tactics. It is merely the artificial improvement of a natural position, and the very smallness of the trenches lessens their claim to be considered as feats of engineering.

Shock action involved concentration; short ranges and smoky powder rendered concealment abortive, so that large upstanding parapets, enclosing a small space, protected at once a mass of infantry, guns, and stores, etc.

But these old ideas of large, solidly constructed forts and redoubts in the field are now obsolete. Works of this nature form conspicuous objects in the distance; concentrated shell fire can be directed on them, and, especially with the high-explosive bursting charges now used, they would form veritable death-traps. The Boer forts round Mafeking, mostly built on old models, had to be divided into regular compartments by sandbag walls, in order to mitigate the effect of even the puny 7-pounder shells which might burst in them. Even converging rifle fire directed on such a target would sweep the work. When fired from a distance the bullets would fall at such an angle as to be practically impossible to provide against. While involving a great amount of labour such works afford but little real safety.

It soon became realized that a low-lying parapet gave quite as good cover to the men in a trench immediately behind it, while it was easier to construct, and could readily be made practically invisible at a few hundred yards off. To enclose a small interior space was but to form a useless shell-trap.

It is no more obligatory to mix up guns and waggons with riflemen in fortifications than it is in the open field. So that now, guns in their own emplacements will be posted away on the higher ground behind, while the infantry will be in a chain of numerous widely extended trenches well to the front. A line of simple trenches, then, preferably with parapets on both sides of them to provide against attack from the rear as well as front, became the recognized type of fortification.

There was an interesting instance of the evolution of the trench from the redoubt near Modder River. For the protection of the standing camp there, a number of works were constructed on the hills around. One of these was in the form of a redoubt, a strong parapet enclosing a considerable area of ground. But this was made on a hillside, and when finished, it was seen that the interior of the work was visible and liable to be swept by fire from outside. A large "parados" had then to be constructed across the upper side of the work to protect the backs of the men lining that face. This completed, it became evident that these men might be able to fire over this parados, and form a supporting line to the men lining the lower face. But as two tiers of fire were hardly necessary, and the enclosed space found to be useless, the one line of men facing either way was considered sufficient, and, in fact, the work practically became a long trench with a parapet each side of it.

The ruling features of modern tactics are dispersion and concealment. Invisibility has now assumed a greater importance, since with no puffs of smoke to indicate the exact position of troops and of guns, concealment from view is comparatively easy. The great art is, consequently, to arrange trenches to similarize with the background; to spread them over the country so as to utilize every tactical advantage of the ground, and to so dispose them as to form no target for the hail of bullets and powerful projectiles discharged against them. *Invisibility of works.*

In fact, under modern fire, it becomes necessary to extend fortifications, just as it becomes necessary to extend individual men.

The old idea of a big thick parapet was to make it impenetrable by common shell. With smaller works and greater ranges this becomes unimportant, considering what an extremely small target is presented. No parapet need rise above 3 feet from the ground (usually much less). At the closest range to which artillery is likely to approach a trench occupied with riflemen—about 3000 yards—the target presented would only appear the same size as a line one inch thick at 100 yards off. This can be better realized by imagining a man at that distance holding out his thumb, which, it must be granted, would not run a great risk of being hit by any one firing at it. Considering, moreover, the insignificance of the result if a shell *did* penetrate, gunners would not find it worth while endeavouring to hit the parapet, and would prefer bursting shrapnel in front on the chance of getting some of the bullets home.

It is, therefore, of great importance to make all works as inconspicuous as possible. By covering the front with grass, leaves, sticks, or even *dry* earth (which is often differently coloured to the damp, newly dug earth), they may soon be made practically invisible at over 800 yards. Artillery, and even riflemen, will then have no mark to guide them, and much of their fire will be wasted.

If within sight of the enemy's scouts, the front of trenches should be covered as much as possible even while being completed. At Magersfontein, after the great fight, the Boers were daily occupied throwing up fresh trenches, but the newly dug earth was so conspicuous that, watching through telescopes from some miles away, we used to draw careful sketches of the position of their works.

Straight flat-topped parapets will often show up at a distance simply as being unlike anything

BOER SAND-BAG FORT : MAFEKING.

natural. If they be undulating and irregular they will catch the eye less.

The first consideration for the defence of a position is, naturally, the selection of the exact site of the works. The tactical position is, of course, all-important, but not only should the general features of the ground be suitable for fortifying, but details, such as the nature of the soil, must be taken into account; other factors also affect the question. Trenches cannot be dug in rock or without spades. Obstacles cannot be formed without wire or bushes, nor can clearings be made without tools (or matches and dryness). Much depends on time, labour, and tools available. *Selection of site.*

Situations sometimes occur when, from the rocky nature of the ground or other causes, it is impossible to dig trenches or where loose stones are not plentiful enough to erect sangars. If it be, nevertheless, of importance to make a work on that particular spot, sandbags may be carried, ready filled, to the spot, or they may there be filled with such sand and bits of rock as can be scraped from the crevices.

At Spion Kop it was found necessary to carry up filled bags, the ground being so difficult to dig. At Cannon Kopje, one of the forts round Mafeking, the soil was so unsuitable for excavations that earth and filled sandbags had to be brought from a distance to construct the works.

Positions may often have to be taken up, such as to guard some particular place, which are not at all suitable to tactical requirements. But though the selection of ground may be most important when artificial improvement cannot be carried out, yet, after all, almost *any* position can be made very strong after a few hours' judicious work. Even if commanded on all sides, if plenty of cover for the attack, if good positions exist for the hostile artillery, yet, provided facilities exist, with carefully devised shelter trenches and gun pits, a proper clearing in front, well-placed obstacles, and covered communications, a naturally quite untenable position may soon be so improved as to be well-nigh impregnable.

Means available. From a practical point of view, one of the most important items for consideration in designing a trench is the *tools* that are available. Again and again it occurred during the recent campaign that, when ordered to fortify a position, it was found that but a very small supply of picks and shovels was to be had. Tools rapidly became lost and broken, and regimental supplies were therefore very limited, and detachments had to do as best they could. In South Africa loose stones were so plentiful that defences could very generally be formed without tools. In European warfare the spade will be a necessity.

Even the physical state of the troops may affect the question, when fortifications have to be constructed on the spur of the moment, and when the men, perhaps after a long day's marching, are by no means keen about the necessary labour. An excellent arrangement was instituted in some of our forces, in having a small corps of native labourers provided with suitable tools always ready, while on the march, to improve roads, and in camp, to construct shelter trenches and other defences.

It is practically impossible to select a good site during the dark, or even in a fog. One has frequently had to take up an outpost position after dark, and almost invariably, when dawn broke, one was surprised to find how indifferent a position it was. Selection in dark or fog.

A trooper of Thorneycroft's M.I., describing the fight on Spion Kop (in the *Empire Review*), says, "With the light came the disappointing knowledge that, from where I was, at any rate—the direct front—nothing could be seen. I had built my bit of wall with great care, with the most delightful loophole, out of which to see and fire, but on looking through it there was nothing except the 50 yards or so of level ground to the edge of the hill."

Having determined upon the site to be fortified, and the general arrangement of the works, the next question to be decided is the exact position and the details of construction. Details of trench.

The rifleman's trench is the all-important feature of modern fortification. Gun-pits, horse-pits, bomb proofs, etc., are merely subsidiaries away in rear.

Trenches are bound to vary greatly in pattern,

according to the nature of the soil, tools and labour available, and the time at disposal.

Field intrenchments may almost be divided into two classes—

First, there is the rapidly constructed cover, which may be scraped up, for instance, during the actual progress of an engagement, whether for defence or for temporary shelter during an attack. Much the same kind of thing will be required to protect outposts put out for a night. Usually such works are quite independent of others, and are merely to form cover for a few individual men.

Secondly, there are the works of a more permanent nature for putting round a standing camp or post.

No very hard-and-fast line can be drawn between the two, and the most temporary shelter may later be improved to any extent.

The Boers would sometimes carry a couple of large stones and lay them down where desired, and fire between them; a poor form of cover, yet undoubtedly better than nothing at all.

Stones, if available, may be piled up without the use of tools and give good cover, more rapidly and easily than can be obtained by digging. But by excavations better cover is gained. In loose sandy soils cover can be scraped up with any rough implement, and a hollow formed in which to lie. So with shingle and small stones a parapet can be heaped up without the use of any tools.

The most usual and most natural form of

temporary cover is to scrape with a spade, or even bayonet, a hollow in the ground about 6 to 8 inches deep, and of variable size, about 1½ to 2 feet wide, and perhaps 5 or 6 feet long, the excavated earth heaped to the front, and some form of head-cover arranged, if possible. It must be borne in mind that on flat ground, a bullet skimming over a low parapet may catch the heels of a lying man.

It is important to see (1) that the firer's head and shoulders are covered from the front, and, as far as possible, from the sides. (2) That he is able to *see* and fire on the ground over which an enemy may be expected. (3) That the works are as inconspicuous as possible.

Men on outpost will often dig a ditch in front of the parapet instead of a trench in rear, thereby not having to lie on damp or newly dug earth. They *will* study their own comfort before efficiency, but by doing this they only get half as much cover for the same amount of labour, and they are not bound to lie in the trench till firing begins.

The parapet thus erected is often hardly of sufficient thickness to be bullet-proof. Still it may form cover from view, so that the enemy does not see where to fire back, and it gives a man *confidence* to be behind ever so slight a cover.

Such "pits" will often be so close together as to practically form a trench. If the parapets join in continuous line better cover is got, especially from oblique fire.

FORTIFICATION.

It will, in most cases, be best to commence a trench such as will give useful cover at once, and which can be improved as time goes on. Thus, though a narrow and deep trench, for standing, was usually considered the most desirable, such a pattern was of little use until actually completed, which involved a considerable amount of digging.

Perhaps the best design, as giving immediate cover, with capability of subsequent improvement, is a chain of diagonal pits or excavations for lying in, about 2 feet apart, roughly about 5 feet by

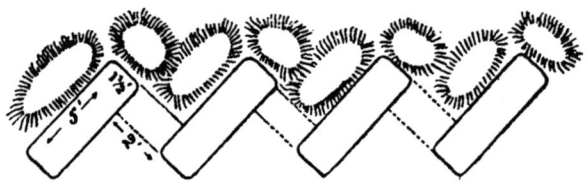

DIAGONAL SHELTER-PITS.

1½, the direction being about 45° to the line of front, the excavated earth thrown up to the front and left front. As for depth, 6 to 9 inches gives quite good cover, and the deeper they are made (rear end first) the better, until, connected up, they become a serpentine or zigzag standing trench. With such a form good cover can be got with an excavation of 5 cubic feet per man (other forms mostly require 10 to 20 cubic feet). It will be found quite practicable to fire 45°, and even 90°, to the *left* from the line of the trench in lying-down position.

FORTIFICATION.

One of the first considerations in designing works for the defence of a position is to decide on the length of trench. The Boers, notably on Laing's Nek, constructed long, continuous trenches stretching for miles on end. From this form to rifle-pits to hold one or two men there are many varieties. *(Long trenches v. short.)*

In order to present the smallest target, men should be as widely extended as possible. If a number be crowded together into one small work, the reports of the rifles and other signs will inevitably draw fire to it, and out of a large number of bullets a certain percentage will get home. If a shell happens to burst in or close over it, the whole garrison may be exterminated.

But if, on the other hand, the men be extended, each in his own pit, 20 yards apart, there can be no concentration of fire, and a shell could only harm one man.

A continuous trench, however, gives certain advantages—

1. Control by the commander is often most necessary to direct the fire on particular targets, and prevent the mere waste of ammunition.
2. Communication so as to pass on orders, to supply extra ammunition, or food or water.
3. Care and removal of wounded.
4. The mutual confidence gained by the proximity of companions.

If time be not pressing and labour unlimited, very long trenches may be made with loopholes at wide intervals. These have the advantage that it would be difficult for the enemy to estimate the strength of the garrison, and his fire will be spread over a wide area, and therefore much of it harmless. If the parapet be struck by a chance shell, the efficiency of the work is but little affected. Space is then available to put reinforcements, and a covered way is formed for communication.

A chain of small trenches may be later connected, and the same result obtained.

Many small posts, such as railway stations, in South Africa, were surrounded by such a long, continuous trench.

If it be desired to hold a given length of front as strongly as possible, the defenders must be almost shoulder to shoulder, and therefore a continuous trench would be necessary.

So that, for a regular position which it is required to hold in strength, a long, continuous trench is perhaps best, and it may even be desirable (as at Magersfontein) to arrange several rows one behind another.

The chief objection to a long, continuous trench is the great amount of labour necessary for its construction.

If rifle pits be placed very close together, some of the above-named *desiderata* can be complied with, but then a little extra labour might connect them and give the full advantages of a continuous trench. The most usual form is a go-between—that is to say, a number of short detached trenches, each to hold about a dozen men.

FORTIFICATION.

It is of great importance in arranging a chain of works, between which an enemy may penetrate, to make each independent of the others, so that if one be captured by the enemy, it does not involve the fall of others. Cover should therefore be provided from fire on the flanks, and even rear, as well as from the front, even though the camp or other works may be behind. It may in any case be desirable to have cover behind, so as to prevent stray shots from one's own side coming in at the back.

Independence of works.

On Monument Hill near Belfast, a number of small works were placed around the camp occupied by two companies Royal Irish Regiment. A thick fog prevailed one night when, at about eleven o'clock, a determined attack was delivered by a large force of Boers. It is said that the enemy, rushing up from all sides, succeeded in penetrating between the detached works, and then fired back into the rear of the works, which had no protective parapet behind. This led to the whole hill being taken. It would have been practically impossible under such circumstances to have made a continuous trench right round the hill, and if it had been made, it would have been so weakly occupied, that parts of it would probably have been rushed. But had the works been so constructed as to have all-round protection and all-round fire, it is improbable that the Boers could have captured all the surrounding works, and would then have had to beat a retreat under a close and galling fire. At

Nooitgedacht, when the Northumberland Fusiliers held a chain of outposts, the Boers managed to slip through, and fired into the backs of the works.

All works should then, whenever possible, possess both cover from all sides, and also enable the garrison to fire in every direction, whatever the form of the trench may be.

It should also be arranged, if possible (as it generally is), that the defenders can fire in *all* directions, without much chance of damaging their fellow-defenders. In the usual undulating or broken ground protruding features may be arranged to intervene. On flat or open ground, a small bank or wall may be specially erected between two works (see figure on p. 93).

Position of trenches. The position of a trench is often on a hillside or rising ground, in which case it should not be on the crest of the hill, whether this be the top or merely a protruding under feature, for the reason that it would then stand out against the sky-line to an attacker. The conspicuousness of a trench so placed, as compared to one a few feet lower is most marked.

If, however, for other reasons a trench has to be made in such a position, by careful arrangement, such as undulating the parapet, planting bushes on it, arranging heaps behind loopholes so that the light does not show through, this objection may be modified.

A mistake often noticeable is the placing of a trench so far back on a convex hill that the bottom

FORTIFICATION.

of the slope is unseen. An officer standing up, especially if mounted, may choose the site as a good one, without noticing the dead ground in front. After the trench is finished, it may be discovered that the rifleman, peering through the loopholes, is unable to see, perhaps, more than a few yards of ground to his front.

A gentle and practically unnoticeable rise of ground in front may have the same effect. It is, therefore, most necessary, before deciding on the

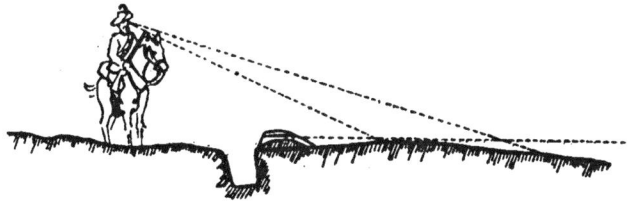

VIEW FROM LEVEL OF TRENCH.

exact position of a trench, to place the eye on a level of where the loophole is intended to come.

But there is no necessity for a trench to command a distant view. As has been said about positions, so long as a rifleman has a fair field of fire, it is not necessary for him to see further. So a trench placed on absolutely flat ground (provided that drainage is allowed for) may be in a most excellent position. At Colenso the Boers had a trench some 500 yards long extending across an almost flat plain (with a slight slope towards the river). This was in a most *unexpected* situation, out in the open halfway between the river and the hills behind.

It would have formed a most appalling obstacle to Hart's brigade had they got across the river.

Trace. As regards the "trace" or plan of works, the first consideration is *not* to dig the trench in a straight line. The chief reason for this is to prevent the possibility of its being enfiladed. Sometimes one has seen an officer carefully noting the line of prolongation of his straight trench, saying it is all right because that prolongation does not rest on ground capable of occupation by the enemy. But it has to be remembered that bullets, shells, and even shrapnel bullets, if coming from *nearly* in line with the trench, may ricochet off the sides and glance along the trench. Also that chance shots, fired, perhaps, from downhill out of sight, will, if they happen to be in line, rake the trench and probably hit two or three men.

Another reason against the construction of a simple straight-line trench, is that the field of fire is limited to the immediate front. If the enemy attack such a work on the flank, no fire can be brought to bear on him. Nor would the garrison of such a trench be able to assist so well in repelling an attack on neighbouring works.

The Boer trenches, though much lauded in some quarters, possessed but one feature well worth adopting, and that is the zig-zag, or winding trace for long trenches, which prevents, to some extent, enfilading, and allows of a much wider field of fire. Also (what was probably the real origin of the design) no time need be lost in "dressing" the line before digging, while difficult spots to dig can be avoided and natural conformations of ground easier followed.

For the small detached trenches the most usual form was to throw back the flanks and make it of a roughly semicircular form. Variations were often used, such as a redan or blunted redan, or even a figure-of-three trace.

But if head-cover is a necessity, it is evident that the trace of an "all-round" work will be very dependent on the form of loophole. The splay of a loophole is very important. If, as so often made, it is only capable of allowing the rifle to be fired through a small angle, a great number of loopholes will be required to each rifle to admit of firing in every direction. A good loophole may be made to splay 45° on each side of the central line of fire— that is, through a quarter of a circle. So that, even then, four loopholes must be constructed for each rifle if ability to fire in all directions be required. This is a point very often lost sight of. Men do not like the labour of constructing four loopholes for each one of the garrison. *Loopholes affect trace.*

If the work consist of a trench with a loopholed parapet to the front and rear, it is evident that the fire will be much restricted, unless care be taken to make it of a suitable trace. With the simple redan, not only will there be "dead" sectors of ground in prolongation of each face, but there will also be a large sector some distance in rear of the work, larger in proportion as the splay of the loopholes and the angle of the trace is reduced. If the work has three faces, as the "blunted" redan, we do not improve away the sectors at the ends. But sometimes such an arrangement may be of

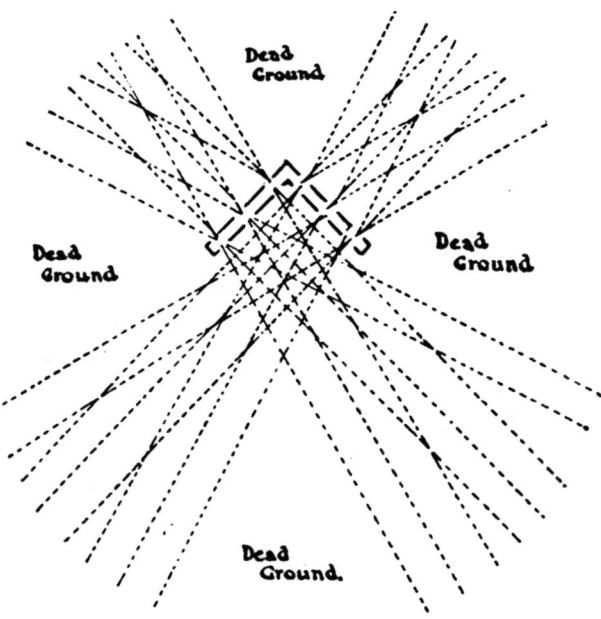

NARROW LOOPHOLES LIMIT FIRE.

advantage, where there is a chain of works, to prevent the garrisons firing into their neighbours.

The subject is one well worth careful consideration, for, without doubt, the cause of many a mishap may be attributed to the want of forethought in the construction of works, and large portions of "dead ground" being left unswept by fire. Badly arranged loopholes limit, not only the field of fire, but of vision also, and if the enemy come up on all sides of such a work, those approaching over the dead sectors will be able to get right up to the work.

It is not sufficient to arrange that every portion of the ground is covered by *one* loophole, since if the attack come from this direction, the one man occupying it may be hit, or it may become knocked in, and the advance continued with absolute impunity. It is certainly desirable, whenever possible, to so arrange a work that every man of the garrison is able to fire in any direction. It will be found that to accomplish this the work must theoretically be of circular or polygonal plan. If the loopholes admit of fire through a right angle, a four-sided work is necessary to avoid dead ground, but if they, as so often constructed, only permit a rifle to be pointed to about 15° on either side of the front, a greater number of faces will be required. But then, if the enemy gets all round such an inclosed work, bullets will be liable to pass over the parapet and take in reverse those lining the trench on the far side (unless the parapet be very high). To erect an inner parapet, or parados, would seem a useless amount of extra labour.

Probably for practical purposes a semicircular trench is best. There is then, at all events, no completely dead ground, and if the trench be of

L

162   FORTIFICATION.

such a length as to provide four loopholes for every man, practically "all-round" fire is obtained.

Head-cover. In order that the cover may be really efficient, it is imperative to provide head-cover—that is, to build up some form of loophole.

The great objection to the construction of such head cover is not only that extra labour is necessarily involved, but also that badly made loopholes,

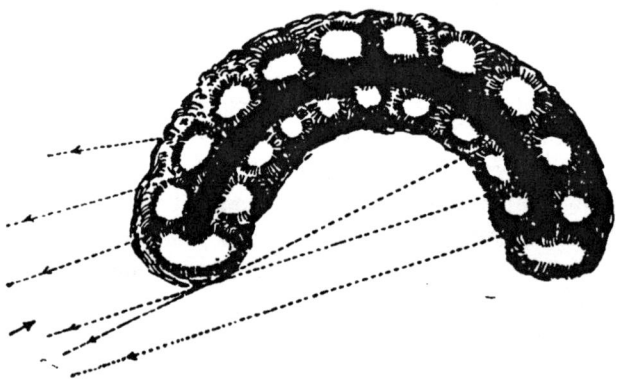

SEMICIRCULAR ALL-ROUND TRENCH FOR SIX MEN.

and even those carefully arranged, must limit to a greater or less extent, both the field of fire and the view. The construction of good head cover is therefore a study of great importance. The old regulation flat-topped parapet is, under most circumstances, *worse* than useless. To fire over it, a man must expose his head and shoulders; and if there be a heavy fire going on, and a shower of bullets spattering all about the parapet, with many a "whiz" and a shriek, the average man will *not* put his head up; it is not good enough. He will certainly duck down and keep well under cover, and then the enemy can approach with impunity.

I say "*worse* than useless" for this reason. Such a trench is usually conspicuous (unless most carefully made up); it then forms a target or guide to fire at. A hot fire will, in consequence, be brought to bear on the trench, making it highly dangerous for a man to expose his head above the parapet. If he remain crouching behind such cover, he is not in a position to fire back. Now, if the man, instead, be lying flat, out in the open, he will scarcely be visible except at close quarters. He will generally be able to gain some natural cover, at all events from view. He lies then practically unseen, and draws no great amount of fire on the place. Yet he is able to fire back in the direction of the enemy, even without raising his head; so that he is really better off without the parapet in front of him.

Head-cover not only protects a man from being shot, it also gives him confidence, and thereby enables him to shoot better. The enemy, too, not seeing heads popping up here and there, is unable to ascertain if a trench is occupied.

It is not only direct fire that must be guarded against. Ricochets, striking the top of the parapet or the ground near, often turn off at very sharp angles, and are still capable of doing deadly harm.

Shrapnel bullets and bullets fired from a long way off will also fall at a steep angle. A Lee-

Metford bullet fired at a range of a mile will descend at an angle of 1 in 10. Shrapnel bullets from a distance of 4000 yards fall at about 1 in 4.

The inside slope of parapets should, for this reason, be built as steep and as close to the trench as possible, so that a man can lie close under it. Men are often apt to heap the earth well to the front and build their loopholes on top in such a way that when firing through them, the man's head is, perhaps, a couple of feet behind it.

Thus a bullet, especially from shrapnel, coming

HEAD-COVER.

within a few inches of the top of the parapet will just catch the man's head. Also the closer the eye is to the loophole the more extended will be the view. By reveting the parapet, the loophole can be constructed close over the trench, and the parapet even made to overhang, so that the man's head is better protected.

When firing through a loophole the rifle rests on the parapet, and it is quite unnecessary to hold it with the left hand.

Loopholes. As regards the *size* of a loophole, there can be no doubt that it should be as *small* as will possibly admit of efficient firing and seeing through. This

SPRING-GUN ON FENCE.

GOOD LOOPHOLES (PIET RETIEF).

involves a certain size. Yet it is a very common fault to make them too large, on the supposition that a better view is obtainable. If the opening be larger than a man's head (as is often done), no cover is practically afforded, at all events to direct frontal fire, and he might just as well have his head above the parapet. Not only is the loophole easily seen from a distance, but with a hail of fire a large opening will let in a good proportion of the bullets. One is apt to forget the enormous mass of bullets that can now be poured on a target. It would be nothing very extraordinary for a thousand men to fire fifty rounds apiece at one object, such as a conspicuous trench. If the great majority of these shots struck within a space represented by a target as large as 50 feet long and 6 feet high, it would mean that on an average there would be one bullet on *every square inch of the target!* and they would be thicker towards the centre. Such a result may be exceptional, but these figures enable one to realize what it means to face modern musketry, and shows the importance of very carefully considering the subject of the construction of cover. Every square inch of human flesh left exposed is liable to have a bullet put through it.

The actual opening of a loophole (as in a steel shield) need be no larger than 2 by 3 inches. For view a horizontal slit $\frac{1}{4}$ inch high is sufficient. But they are often made as big as 6 by 10, which would increase the chance of being hit by ten times.

With bullets pouring like so much rain, it becomes

solely a matter of exposed vulnerable area. Now, a man lying down presents a target of at least 200 square inches; but if a loophole give a surface of, say, only 10 square inches, through which a bullet can pass, it is clear that if both sides are firing an equal number of bullets, that which is intrenched with good loopholes, should only suffer one-twentieth the loss of that lying down in the open, and if the latter rise to advance, the proportion will be much greater.

This proves the supreme importance of intrenchments with head cover, and shows that every square inch of cover is of consequence.

But though it may be of great importance to make the loopholes very small, we must not thereby cripple the efficiency of the musketry fire. If the loophole consist of but a tube through the parapet (as practised by some native tribes), fire can be delivered only in the one direction, and the firer can see nothing of what is going on, except within the very limited area visible.

It is important that a man in an intrenchment should be able to obtain a good view of the ground in front of him. If the view be very limited and he sees nothing to his immediate front, he may be inclined to raise his head to peep over the parapet and see what is going on. This may not only be dangerous for the man, but may "give away" the position.

The best shape will therefore be wide and low. A loophole need never be more than 3, or at most 4, inches high (the width of the fist).

FORTIFICATION. 167

With a little care, loopholes, combining all the desirable qualifications, can be made, and it forms a most desirable subject in which to give full instruction to the private soldier.

Construction of loopholes.

The construction of good loopholes is very dependent on the materials available. With square-shaped stones, often to be got in South Africa, very effective head-cover can be built up. But if nothing more than loose earth is obtainable, a different sort of arrangement must be adopted. In the latter case, or if time be pressing, castellations or embrasures is all that can be attempted; but if carefully made, even these may be very effective—

NOTCHED PARAPET.

at all events giving protection from bullets coming from the right or left front. As in such a case the *height* of the aperture cannot be limited, the *width* must be as narrow as possible, in order to reduce the area to the smallest extent. Then only a vertical strip of the head, perhaps only 2 inches wide, is visible from the front.

The sides of such an embrasure should be as steep as possible, and at least 6 inches high, to give good protection.

Loopholes, in order to allow lateral range of fire, must be splayed. There are three methods of accomplishing this—

168   FORTIFICATION.

(1) Narrow outside and wide inside (like the letter A). The chief merit of this form is that a very small mark is exposed from outside, either to view or penetration. The sight of a small loophole, it is true, is usually quite unimportant, as it could never be seen for more than a couple of hundred yards.

But there are exceptions. For instance, at Mafeking, where the trenches were sapped out to within

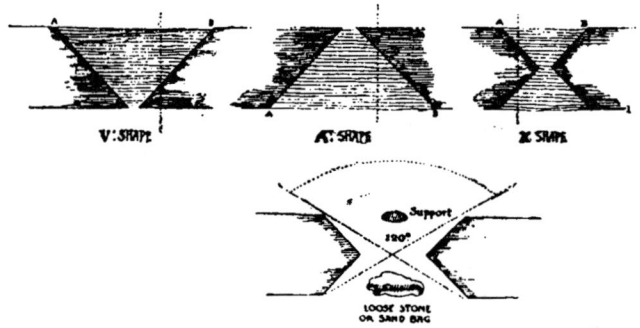

PLAN OF LOOPHOLES.

70 yards of one another, accurate aim was taken at the loopholes, and the defenders scored greatly by introducing steel plates with a small loophole, while the Boers only had the clumsy sandbag arrangement.

(2) Wide inside and narrow outside (letter V). This enables the defender, by putting his eye close to the hole, to obtain a much better view, if the angle be wide. Such a shape is often said to form

a funnel to deflect bullets inwards, though with soft earth they do not ricochet. It may, however, be a little more conspicuous.

(3) Wide both outside and in, and narrow in the middle (letter X). This, while possessing in part the qualities of each of the above, has the great advantage of giving a thicker parapet to oppose the bullets at the narrowest part, and also, not being so wide across (at A B) as the others, is easier to bridge over with stones, turf, etc.

On the whole, for a roughly notched parapet, the form of No. (3) is best, since, with an embrasure, view is not so important, it being easy to see by slightly raising the head, and strength of parapet is important.

Such an embrasure may later on be improved into a good loophole if a board, sheet of iron, large flat stone, or piece of hurdle-work be laid on top of it and covered with earth.

The earth above a loophole should be at least 6 inches high to give protection. A greater height is desirable, but the higher the parapet stands, the more conspicuous, and the more liable to damage is it likely to be. It must also be thick enough to be bullet-proof, a point often forgotten.

But if materials, such as above mentioned, are available, it will usually be preferable to go in for a rather different design.

In constructing a loophole, it is of course most desirable to give it as wide a splay as possible. But to get this, good materials must be procurable.

If the parapet be 2 feet thick, and if supports can be got as long as 4 feet, a splay of 120°, or one-third of a circle, can be obtained; but it is seldom that flat stones could be got long enough; sandbags would not span this, and even brushwood would be liable to sag. So that really good loopholes in earthwork are dependent on boards, poles, hurdles, or corrugated iron being obtainable.

The private soldier is very fond of piling up a few small stones in front of him, which, if struck by a bullet, would fall like a pack of cards. He also often puts up bits of turf, offering practically no resistance to a bullet which can penetrate two or three feet of earth. Men will *not* understand this until they have awkward experiences of what a bullet can do. An officer inspecting a work should go along with a stick and knock down all he can, explaining that a bullet will come with far greater force than that which he can put in with his stick. Men should be practised in peace time constructing trenches on a rifle range, and afterwards firing at them to see the result.

The ordinary sandbag loophole is a clumsy arrangement, and usually leaves much too big an aperture. It is desirable to fill up such with earth to within 3 inches of the top.

Sandbags of the ordinary coloured sackcloth, especially after being bleached in the sun, show out very conspicuously from a long way off, unless carefully covered or coloured with mud, etc.

Loopholes in an earthen parapet should, as a

FORTIFICATION. 171

rule, be not less than 3 feet apart to ensure the parapet being bullet-proof to oblique fire.

For carefully made loopholes in semi-permanent and permanent works, the best form, theoretically, will be very wide and low outside, narrow and high inside. Such a pattern need only be about 2 inches high and 3 feet wide outside, and inside 3 inches wide by 6 inches high. This gives a good range of vision and fire, both laterally and vertically, and yet presents the minimum of target to penetration. The vulnerable target presented is then only 2 by 3 inches.

It is always desirable to have a large stone, or bag of gravel, inside a loophole, to shift as required.

There should be at least one loophole for each man of the garrison on *each* face of the work, remembering that some may be filled up or damaged by fire, and officers and others will require look-out places in addition to the actual riflemen. *Number of loopholes necessary.*

In many of the block-houses erected during the later phases of the war, only two or three loopholes were provided to each side. If attacked from that side, the remainder of the garrison (usually seven in all) could do nothing but wait, and could not even so much as watch the enemy's approach.

Though the oft-mooted idea of a soldier carrying with him a bullet-proof shield appears to be quite impracticable, yet a steel plate with neat loopholes in the centre, such as were used at Mafeking and in some of the block-houses, proved to be of the greatest value in constructing semi-permanent works.

Excavations and erections. The requisites of a good work are to give ample cover while admitting of full facilities for firing, and being as inconspicuous as possible. Cover can be obtained by excavating the soil, or by erecting a parapet, or by both combined. Generally speaking, the deeper the trench, the better the cover; the higher the parapet, the more visible it becomes. This might seem to imply that wholly excavated works are the most desirable. But though practically invisible, they are liable to be flooded by rain, and, unless the site be specially favourable, the field of fire is likely to be obstructed by irregularities in the ground or herbage.

Some of the Boer trenches were of this form—that is to say, without parapets—the excavated earth being scattered around or removed. Though there was then absolutely no guide to the eye to distinguish the work when unoccupied, there was an absence of head-cover; so that when the heads and shoulders of the firers popped up, it made the position visible. Besides this, double the amount of labour was necessary for the same amount of cover as would be obtainable had the earth been heaped to the front. Some parapet, though it only be cover for the head, is desirable, and if the ground be slightly sloping to the front, so as to get a good view of the field of fire, the rifle may be fired from the ground line, with a parapet above only 6 or 8 inches high.

But, on the other hand, in flat, low-lying country liable to flood, or when long grass or rough broken surface prevails, it may be necessary even to build up a high parapet wholly above ground. Such, too, may be necessary on hard rocky soils where digging is impracticable. "Schanzes," or walls of stone, were often built up, a second wall being usually added for protection from reverse fire. At Mafeking one of the forts on Cannon Kopje had to be constructed mostly of sandbags and earth brought from a distance.

But "erected" works usually possess in a greater or less degree the objection of being visible from a distance. To be inconspicuous, a parapet should be made with a very gentle slope to the front. However much "similarized" to the surrounding ground, even if turfed over in a grassy country, a steep bank will always stand out, while a very gradual slope will scarcely be distinguishable.

A high parapet should therefore only be constructed either to gain command, to avoid flood, or on account of difficulty of digging.

Excavations usually give better cover than embankments, since the interior face may be steeper. With an excavation it may be perpendicular or even overhanging, while an unrevetted heap of earth will seldom stand more perpendicularly than 1 in 1. The further the man is behind the crest of the parapet, the more danger does he run of being hit by dropping bullets.

Section of trench. In South Africa deep trenches found much favour. One novel feature of the Boer trenches was the manner in which a deep but narrow trench was *widened* at the bottom. This had the advantage of making them much more roomy and convenient without detracting from the cover gained, and with less labour than a wider trench would involve. But this is purely a matter of convenience, and with a loose soil such overhanging sides would be very apt to fall in.

As regards width of trench, while formerly room for lying and for kneeling was usually allowed for, a standing trench was most used in semi-permanent works in South Africa. The latter, though necessitating digging to a depth of at least 3 feet (usually 4) need be but little over 1 foot wide, or half that necessary for kneeling, so that as regards volume of earth excavated no greater labour is required, while better cover is obtained. The kneeling position may be good enough for a few shots on the range, but is very irksome when crouching behind a parapet for hours on end.

On the other hand, the earth deep down often becomes hard and rocky, and such trenches are more difficult to drain.

Covered-in trenches. In considering works of a semi-permanent nature—that is, when time, labour, and material are plentiful—it becomes advisable, not only to guard against attack in flank and rear, but to make the works, as far as possible, absolutely impregnable against assault. In order to attain this, it is desirable to roof in the trenches so as to get overhead protection. Such an arrangement ensures: (1) Full protection from rifle bullets from all sides, whether enfilade, reverse, ricochet, or dropping (except such few as might come through the loopholes). (2) Absolute protection from shrapnel bullets and splinters of shells (whose angle of descent would be so great as not to enter the loopholes). (3) Practical impregnability against assault. The assailant could only capture such a work at enormous loss, and if provided with tools or explosives to demolish it. (4) Cover from rain and exposure, preventing the trench from being flooded and affording shelter to the garrison.

The Mafeking garrison, in making their assault on the Boers in Game Tree Fort (an erection of sandbags) found the frontal fire very intense, and made their way round the flanks and finally got in rear of the work, but found it enclosed and an all-round fire still being poured forth on them. Finally, they rushed the work, several actually climbing the parapet, only to find that the work was roofed in, so the assault was recognized as hopeless.

Later on, General Baden-Powell, with some of these same men under him, utilizing the practical knowledge gained in Mafeking, erected for the defence of Rustenburg and elsewhere a number of deep trenches, roofed in so as to form regular underground block-houses. The trenches were about $3\frac{1}{2}$ feet deep, with very wide and low loopholes (about 4 feet long and 4 inches high) formed by earth supported on boards or sheets of corrugated iron, and the whole roofed in with the same materials and covered with a foot or two of earth on the top, the outside being turfed over to match the surrounding country. A narrow winding entrance was arranged at one end.

Such trenches were much used afterwards at Zuurfontein, Heidelburg, and elsewhere, and in fact became the official pattern of block-house for the South African Constabulary; some, on this principle, being constructed to comfortably house comparatively large garrisons.

These trenches remained perfectly dry after heavy rains which completely flooded open trenches.

We tried a simpler pattern, both at Edenburg and at Holnek, of roofing in a lying-down trench with sheets of corrugated iron (a material usually plentiful in South Africa from the roofs of houses and sheds), and covering it with earth.

If the trench be required to face one way only, a simple lean-to roof resting on the parapet is all that is necessary; but a good all-round work is constructed as follows:—A trench of 7 feet wide and from 6 inches to 1 foot deep is dug. A low, notched parapet is arranged on each side, and sheets of corrugated iron laid flat on this parapet to form the top of the loopholes. Other sheets of iron about 10 feet long are then placed across to bridge the trench, their ends (2 feet from one end, 1 foot from the other) being bent down and supported on the inner edges of the horizontal sheets. The excavated

COVERED-IN TRENCH, IN COURSE OF CONSTRUCTION.

COVERED-IN TRENCH.

earth (or a lot of loose stones) is then piled against the front and rear, and on the top. A deeper trench may be dug out along the back part of the trench to allow head room. Wires may be stretched over the top and pegged down to keep the roof securely on, and form at the same time an entanglement. Such trenches, given the material, can be very quickly erected with a minimum of digging.

In a continuous trench of such a nature, if one sheet be left out in the centre, the gap forms an

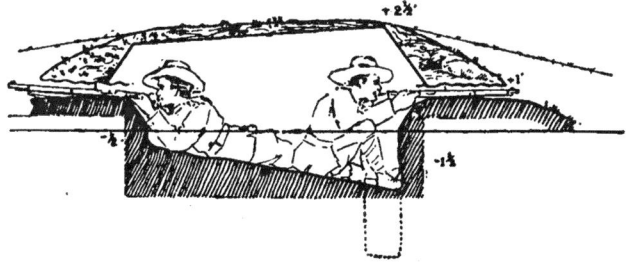

COVERED-IN TRENCH.

entrance to the work and is such that an assailant must enter feet first, and can then be bayoneted before he can bring his rifle into play.

Had covered-in works of this nature been constructed from the beginning of the war, many unpleasant surprises of posts which occurred could not have succeeded. A dozen men in such a work could hold out against hundreds, presuming, of course, they were well provided with water, food, and ammunition.

Simple little pits to hold three or four men, roofed in, with loopholes in each direction, were often used by the black "watchers" on the Lines of Communication.

Block-houses. In the last stages of the war, when the Boer artillery had practically ceased to exist, block-houses were erected all over the country. These were used chiefly for the protection of the railway, but were also placed around towns and along certain lines.

The usual dimensions of these were of such a size as to accommodate seven men, of a circular or polygonal plan. They were mostly made of corrugated iron of double thickness, with a space of about 4 inches between the sheets, which space was filled with stones broken down to pieces about the size of lumps of sugar. A galvanized iron gable roof was added.

About a dozen loopholes were cut, of the **X** pattern, so as to give a good lateral splay.

Some larger block-houses were built up of solid stone in two or three storeys with steel loopholes. These, if more expensive to construct, were much more efficient.

All block-houses were surrounded with elaborate wire entanglements.

These formed healthy and comfortable barracks for their little garrisons, but they possessed several great objections from a military point of view. First, in the earlier pattern at all events, there were too few loopholes. As a rule, only two pointed in any one direction, so that during an

TYPICAL BLOCK-HOUSE.

TYPICAL BLOCK-HOUSE.

attack only two men could fire. Secondly, they would be untenable directly a gun of any sort was brought against them. Thirdly, they formed so conspicuous a target that a concentrated rifle fire could be brought to bear on them, and a certain proportion of the bullets fired would be likely to enter the loopholes. If, say, three hundred men attacked such a block-house, and lying down anywhere within a fairly close range, each of them fired fifty rounds, then, if only one in a thousand took effect, fifteen shots would actually enter the building, and if the garrison were attempting to fire back, they must be struck in the face with these bullets. Other shots penetrating the roof would distribute ricochets and splinters about the inside. Very luckily they were seldom attacked with energy, and the Boers fought rather shy of them.

The doors were generally made so small that only one man at a time could creep in on all fours, but if the garrison happened to be outside when attack came, they would take a long time crawling in one by one, and would mostly be shot down before they could all get inside.

To obviate these objections, some rather different patterns were occasionally erected. A nearly flat roof was not so conspicuous and not so easily penetrated by bullets. If this roof be covered with stones (better than earth, which is liable to be washed away by rain), it may be made bullet-proof, which is specially desirable when the work is commanded by a hill.

FORTIFICATION.

A few shiftable sandbags or large stones inside the loopholes enable the latter to be kept closed till wanted.

The doors should have a wall built in front of them to keep off bullets.

Block-houses, like some Boer forts, were occasionally built up entirely of sandbags.

But, on the whole, the covered-in trenches of the South African Constabulary were probably the best form of block-house, being practically shell-proof, and quite invisible at a distance. Some Boers, captured on the South African Constabulary block-house line in the northern Orange River Colony, asked how it was that they were always being ambushed in that neighbourhood. They said they knew all the block-house lines, and were much surprised when they discovered that there was a permanent line of "ambuscades" there.

Gun-pits. It might seem that gun-pits, not being so likely to be closely approached by an enemy, need not be so inconspicuous as rifle trenches. But when artillery fire begins, telescopes and glasses by the score will be directed on the most likely places, and then, if not well concealed, a gun emplacement may soon be discovered, and become the target for a concentrated fire.

The mere gun itself is a by no means conspicuous object from a point of view in front of it. It is not much more discernible than a man's head. The wheels are more conspicuous, so is the trail. All that can be seen peeping over a parapet is the

FLAT-ROOFED BLOCK-HOUSE.

A LOOK-OUT TOWER.

FORTIFICATION. 181

small muzzle of the gun and the tops of the wheels. Indeed, with some patterns of gun, such as the naval twelve-pounders, the wheels were so small as to be below the gun. At Magersfontein, day after day, the positions of the Boer guns was watched for, but so well concealed were they that their exact whereabouts was not discovered.

The only other objects to hide from view are the men working the gun. These, by moving about, are more easily caught sight of. A screen may be erected, such as a bank of earth with an embrasure in it, and if this can be made inconspicuously, the whereabouts of the gun will be practically impossible to discern. But even with guns firing smokeless powder, the concussion and blast from the muzzle is so great that usually a small cloud of dust is thrown up, especially if there be an earth parapet close in front of the muzzle. The Boers are said to have often provided against this by watering the ground or putting ox hides on the parapet. Where a wall can be built up, the muzzles of long guns can be protruded well over it, and thus not blow up as much dust in front. A few branches or bushes put on the top of the parapet or even on the gun-wheels may hide a good deal. The chief object is to hide the gun and detachment from view; the secondary object, to protect the men from hostile bullets. The latter are not likely to be numerous if the gun is invisible.

A parapet with an embrasure, however, limits the lateral range. A few sandbags, easily shiftable, were often placed on top of a low flat parapet, which gave all the effect of an embrasure, without limiting the direction of fire. Occasionally a work was built up with several regular embrasures in different faces, so as to enable the gun to be fired in any direction.

Since a high built-up parapet is generally conspicuous, and dust is more likely to be raised over newly dug earth, *natural* cover, such as a bank, a large rock, a hedge, etc., is generally preferable ; and as a few hundred yards one way or another is immaterial to the range of the gun, such can usually be utilized to some extent at least.

Old-fashioned gun-pits and epaulements used to be made to face one way only. The most usual pattern used in South Africa consisted of a circular pit, perhaps 1 foot deep, and the earth heaped up all round, the diameter being such as to allow the gun (whatever its size) to be "run up" and back, and slewed round as required to fire in any direction. The parapet was generally reveted inside, if possible, to give better cover.

Gun-pits, as well as rifle-pits, can often be made during an attack. Indeed, as there is usually no immediate hurry for a few minutes for artillery to come into action, it would generally be desirable to scrape up some cover before opening fire and disclosing the position, especially if there is the slightest chance of hostile infantry being, or coming, within rifle range. Had our gunners at Colenso dug cover before coming into action, a very different tale might be tellable, for they could have safely lain in their trenches till relieved or darkness came on.

A machine-gun on a high-wheeled carriage (which, owing to its cumbrousness and conspicuousness, will probably be discarded in future) can be quickly rendered somewhat less conspicuous by sinking the wheels in two narrow trenches. *Machine-gun-pits.*

At Rietfontein, near Pretoria, some small pits, about 3 feet deep, were dug to accommodate tripod Maxims. Head-cover was provided by boards with earth piled on them, supported at each corner of the pit.

When it is necessary to place a whole town in a state of defence, there will usually be three lines. *Defence of towns and villages.*

1. A series of works at such a distance from the town as to prevent, as far as possible, the enemy firing into it. If only a small force is available, and large guns are opposed to it, it is practically impossible to push the posts sufficiently far out. The perimeter of the town of Mafeking was about 2 miles; that of the outlying works was at first over 6 miles, but bit by bit pushed out till over 9 miles round. But even this was of course not far enough to prevent the town from being shelled.

2. If, as is usually the case, the works are a considerable distance apart, an interior and more continuous line must be formed to prevent any rush of the enemy between them.

3. It is also desirable to have yet a third system of defence in the town itself.

184 FORTIFICATION.

When the Boers broke through the outer cordon of defence at Mafeking, and got as far as to capture the old fort, fire was opened on them from interior lines, which resulted in the capture of the assailants. At Rustenburg, works were placed in the centre of the principal streets, so that had the Boers got right into the town, it would have been quite impossible for them to hold it while these works held out.

With a weak garrison, if such interior lines be well planned, it even becomes unnecessary to attempt to keep the enemy out. Let him be encouraged to break through the outer cordon. He is then surrounded in a bad position, and should not be let out again.

Everything will, of course, be dependent on the natural features of the ground, which will modify the disposition.

If the town, as in many instances, lies in a depression with a circle of hills round it, a good chain of posts could be taken up on the hills, so that no gap of over 300 or 400 yards existed between the works, and the enemy may not then be able to see sufficient of the town to bring any considerable fire to bear on it. Here outer forts and inner chain of defence could be combined, while certain places, such as the Ordnance and Commissariat stores, railway station, and head-quarter offices, should be placed in a state of defence in case of the enemy breaking through.

It may here be desirable to say a word about *garrisons*.

Besides the regular defenders occupying the trenches around, there is usually, in a town, quite a large number of men not included as defenders. Telegraphists, storemen, artificers, clerks, and many other trained and armed soldiers are often to be found in various out-of-the-way offices and stores in a town. If each of them has his definite orders for rendezvous, in case of alarm, at certain fortified centres, the garrison may have a much needed addition. There will also always be a certain number of civilians who, even if unarmed, may be made good use of as ammunition carriers, stretcher bearers, diggers of trenches, and even sentries. Every one in the town should have distinct orders as to what to do in case of alarm.

If possible the whole town should be surrounded by an impenetrable fence to stop a rush. At all events the roads leading into the town should be blocked. Waggons were often placed across the roads with barbed wire entwined among the wheels.

When artillery fire can be delivered on the town, not only should all inhabitants construct bomb-proof shelters by their houses, but pits should be dug at intervals along the streets to enable passers-by to take shelter when necessary.

It is, of course, not desirable, when the enemy can bring guns to bear, to rely on houses as defences. Good trenches can usually be dug outside. But at the same time, houses inhabited at night by soldiers, or store-houses, offices, etc., may well be loopholed and prepared for defence against a sudden night attack.

Hedges and walls. Existing hedges, ditches, and even walls, are not often suitable for occupation, and it will generally be preferable to construct new and separate works instead of devoting labour to their improvement. They have not been laid out with any regard to tactical advantages, and will therefore seldom be in the best position for defence. They will usually be conspicuous, and therefore will draw fire, and, if straight, will be liable to be enfiladed. As a rule, much labour would be required to render them efficient. They may, however, form useful obstacles or covered means of communication.

Armoured waggons. Armoured bullet-proof waggons were recently tried with much success. A convoy moving along a road included several of them. If attacked, though the oxen might be shot down, it would be impossible to capture the convoy without taking the men in these movable forts. When in camp, the animals and waggons were parked in the centre, and the armoured waggons put out at the four corners to act as forts.

These were also useful in many other circumstances, such as guarding watering-places if distant from camp; for grazing guards for cattle (getting fresh pasture each day); and for outposts generally. They save the men much labour in digging intrenchments, and so enable posts to be continually shifted to more useful points, which an officer often

AN ARMOURED WAGGON

FORTIFICATION.

hesitates to do in consideration of the extra work involved for the men.

A few important items have to be remembered to add to works.   *Adjuncts to fortifications.*

Drainage of rain-water should be arranged for. On sloping ground a small trench on the uphill side, a foot or two away, may suffice to prevent water running into the work. Many instances occurred of the trenches becoming filled with water after heavy rains. At Poplar Grove, after we had taken the position, the Boer trenches still had muddy pools in them, and must have been well-nigh unoccupiable.

Sanitary and cooking arrangements must be made near works likely to be held during prolonged fighting. All the outlying works at Mafeking had winding trenches, by which the occupants could move away to construct their latrines, refuse-pits, and kitchens without exposing themselves.

The protection of horses, waggons, etc., from hostile fire must be arranged for when a small force has to occupy a position for some time.

If no natural depressions in the ground exist, pits may be dug. At Rustenburg, huge trenches 6 or 7 feet deep were excavated, with ramps, in which to pack the horses in case of bombardment. At Mooi River, where a position was taken up in perfectly open country, pits were dug in which to hide away water and ammunition carts, ruts 18 inches deep being dug at the sides for the wheels, to increase the cover gained.

188   FORTIFICATION.

Communication trenches may be made to connect important works, or groups of works, with the central base. These will be most useful during protracted operations for conveyance of reinforcements and reliefs, supplies of food and ammunition, and removal of sick and wounded. The plan would, of course, be serpentine or zig-zag, depth 3 to 4 feet, with earth thrown up on each side, and occasional loopholes may be put with advantage.

Screens formed of bushes, hurdle-work, canvas, sheet iron, etc., have been much used as cover from view, and may even be used for the above purposes when trenches are a difficulty. The ends of streets at Mafeking, Rustenburg, and other towns were so screened, to prevent snipers at a distance shooting passers-by.

Bullet-proof screens may also be useful. Boards or sheets of corrugated iron fixed a few inches apart, and the space between filled with gravel or broken-up stone, were frequently used. The hospital at Mafeking was protected by such screens. Block-houses were made on this principle, and even armoured trains and waggons were occasionally so protected.

Means of communicating between works and head-quarters should, if possible, be arranged for. Telephones were largely used for this purpose in most of the principal towns in South Africa for connecting the outposts with head-quarters. Electric bells were fitted to the works at Zuurfontein. Speaking-tubes have also been arranged. Even a simple bell-wire is useful to call attention to outlying works, and prearranged signals may be adopted, such as one tug meaning "Are you alert?" to which the sentry may reply with a similar tug, signifying "All's well." Continuous tugging may imply, "Alarm! stand to arms;" or, if sent to head-quarters, "We are being attacked." Such wires were supported on crossed sticks loosely resting on the ground, inclining over when pulled.

It is often desirable to establish some simple means of communicating with distant posts. Smoke signals and beacons were much used by the Boers, and on a few occasions by us.

It is very necessary to arrange means of giving *warning of the approach of persons in the dark*. All old tins (of which there are plenty nowadays in a camp) should be thrown on the ground about 200 yards from the works. Any one passing this zone is almost sure to kick against the tins.

Tins are usually hung on the wire entanglement for alarm purposes, or on a single wire specially erected. These tins should not be, as was so often done, hung conspicuously on the wire, so as to disclose their presence from a distance. They are better hung from the wire, so as to just rest on the ground, and are then less liable to clatter on a windy night. An even better plan is to loosely wrap small pieces of tin around the wire, which make a good noise on being shaken, but are quite inconspicuous.

To give notice of a fence being cut by the

enemy, one wire may be rove through staples or holes in the posts, and, at a point near the work, a heavy stone hung from it above some tins. On the wire being cut through the stone falls.

In the Pretoria forts insulated wires were put up which, when touched, rang electric bells in the forts. Spring alarm guns were placed at intervals along the railway fence, and might with advantage have been more frequently used for outpost work. A simple detonator could easily be devised.

Illuminating the field of fire in case of a night attack is another useful *desideratum* to provide for. In the Pretoria forts electric lights were arranged around, with reflectors behind them, so that in the event of alarm they could be turned on. In some places tar-barrels and flare-lights were placed, and so arranged as to be set light to on the pulling of a string from the work, or by a trip-line.

Dummy works of conspicuous heaps of earth or walls have often been made, away from the real defence, in order to draw off the hostile fire from the defensive works. Dummy guns, too, have often been placed to attract the enemy's attention. At Kaalfontein such a dummy was struck by one of the first shells fired by the enemy in attacking the place.

Look-out posts are very necessary. In flat, open positions this may often be difficult to manage without making an erection likely to draw a concentrated fire on it. A look-out nest was sometimes built in a big thick tree. At other places

FORTIFICATION.

timber towers have been put up, and a sandbag parapet protects the observer from bullets.

Range marks.—It is most desirable, when taking up a position for defence, to ascertain various ranges around it. It is usual to measure the distance to certain well-defined objects, but this plan has the disadvantage, that it is very easy to forget the distance, especially during the excitement of an action. In many posts permanently occupied, it was the custom to have a board with the ranges and directions of conspicuous objects marked on it. But a more preferable method is to place marks— such as biscuit tins, whitewashed stones, posts, etc. —at fixed distances, in various directions, for instance, some at 500 yards, some (say two together, so as to be distinguishable) at 800 yards, some (in groups of three) at 1000 yards.

There is one other point having a close connection with fortification, which is the *housing* of men. In cold, rainy climates this becomes a matter of some importance. Housing of men.

The regulation bell-tent is objectionable for three reasons—

(1) Its whiteness and size make it very conspicuous, thereby not only forming a target, but showing clearly the disposition of troops on a position. (2) It is not arranged to be fired out of, and in the event of sudden alarm, men will take a long time getting out through the door (especially when closed against rain). Numerous instances could be quoted, such as Belfast, Tweefontein,

Moedwil, etc., where troops in tents have been delayed in forming up to resist a night attack. (3) It requires considerable transport.

A small, low-lying hut or shelter, close to, or in, the trench, and of a suitable size, is in many ways much better. Even if not combined with the defensive work (and this is certainly preferable for sanitary and other reasons), good huts can be rapidly constructed of branches, turf, galvanized iron, or whatever materials are available. Even shelters of blankets have been much used, and troops camped for months on end in them. Railway tarpaulins were also much used, supported on waggons, or on any posts available.

Obstacles.

Time and space are closely connected in tactical problems. The great object of having a wide, open space in front of a position is to force the enemy to expose himself for a considerable *time* to the fire of the defenders. If this space be broken and difficult to traverse, that time may be more extended, so that the desired duration may be gained by obstructions to advance as well as by extent of ground. But there is this difference: if the ground *close* to the position be very obstructed, not only is it equivalent to a wider space to be traversed, but the enemy will be delayed under the *closer* and more accurate fire of the defence. Artificial obstacles are therefore preferable to a given distance of open ground, necessitating an equal time to traverse.

A little barbed wire may thus be the means of delaying an enemy under fire as effectually as 100 yards of open ground.

A fence (or abattis) can, moreover, be constructed, given the art, time, and material, which will be absolutely insurmountable, and could only be got over by being destroyed or bridged, which will be dependent on the assailant being provided with suitable tools or materials.

Barbed wire may be considered as an important innovation in modern warfare, and is likely to be largely employed in future wars. In South Africa it happened to be plentifully obtainable and was much used. Fences and entanglements of barbed wire certainly form about the most efficient and most easily set up of obstacles.

In putting up a wire entanglement, the chief consideration is the amount of wire at disposal. It will often be desirable to place an obstacle several hundred yards long in front of a work, yet the requisite amount of wire to form an insurmountable entanglement may not be forthcoming.

It has been a frequent practice to run a simple wire along on small pegs, about a foot above the ground. This is, however, of but little real use except as a tripping wire for horses, and may cause a man to stumble on a very dark night. But in daylight, or even on a fine night, a wire is clearly visible and easily avoided. If the wire be darkened (rusty) it is more difficult to see. In long grass or bushy country it may be more useful, but low entanglements can generally be stepped over or trodden down. If two or three wires can be obtained and placed as a fence, a more useful obstacle results; but with four (or more) a practically unclimbable fence can be made if the wires be stretched tight about a foot apart. By adding stays to the posts, and running wires across them, the obstacles can be made more insurmountable still.

Posts are generally conspicuous, and as it is better to let the enemy come suddenly and unexpectedly on the obstacles, they should be hidden as much as possible. Otherwise the more posts the better, as it strengthens the fence, renders getting *through* more difficult, and, if the wires be cut, only a small opening is made between two posts. A number of very thin posts are less conspicuous than a few big ones. If rough branches, with the leaves on, be used, they may look like natural bushes. But posts *must* be strong, else they are easily knocked down or pulled bodily out.

It has often occurred that, though wire was plentiful, there were no posts available (all wood being required for fuel). Piles of stones or even heaps of earth were then used, or a fair entanglement was made of simple lengths of wire stuck in the earth like large croquet hoops, crossed and recrossed and connected with horizontal wires.

Where there are growing trees, most formidably

impassable barriers can be made by stretching wires across. In the Northern Bush-veld we made perfect cobwebs of wire among the trees.

Dongas and river-beds can be made very difficult to pass along if wires stretch from bank to bank.

The works around Pretoria were provided with one pattern of wire entanglement, which was as follows:—

Posts were to be 15 feet apart. They were 6 feet long, 1½ feet of which is buried underground, making the fence 4½ feet high. Five wires were used to each fence. Two parallel fences were employed 10 feet apart, with wires between. Wire stays were put to each post, and horizontal wires carried along them. Winding entrances were arranged in the entanglement, and these were closed by wires at night.

A somewhat similar pattern of fence was made around the works at Bloemfontein, but the fences were only about 2 to 3 feet high, and so badly carried out, that those going to and from the works frequently stepped over them, even on a dark night, in preference to going round to the entrance.

But an obstacle which fails as an impassable barrier may often be most desirable, since a hesitating enemy, who might at once retire on encountering a formidable obstacle, may be more encouraged to advance. His force then becomes very broken up, and if the attack is driven off, many of the attackers will have to retreat over the obstacles, and will then suffer severe loss.

As so-called obstacles were frequently to be seen utilizing vast quantities of wire, yet easily surmountable, it may be as well to point out what the desirable qualities of a wire entanglement are.

(1) It should be at least 3 feet high, else it may be *stepped over*. (2) Also so arranged (with posts or vertical wires, at least every 3 feet) that the wires cannot be pulled apart and *got through*. (3) The bottom wire to be close to ground so that it cannot be *crawled under*. (4) Posts sloped outwards so that it cannot be *climbed over*. (5) Strong *posts* well set up so that they cannot be *pulled* or *pushed over*, or the fence *trodden* down. (6) As far as possible prevent wire from being *cut* (such as using very thick wire).

To surmount a wire entanglement, if it cannot be either stepped over, jumped over, climbed, got through or crawled under, coats or pieces of cloth, sacking, etc., or bundles of hay may be thrown on it, or the wires may be cut with cutters, or chopped with an axe or sharp bayonet. A file is most useful for this purpose, and may be more easily carried. Often the posts may be pulled or cut down. A number of men seizing the wire and pulling together will often soon pull it down. Boers have even dug out the posts during the night, and in some places the fences were so badly set up that they lifted the wire, posts and all, and held it up while the horses passed underneath. The Boers also sometimes broke down obstacles by driving a mob of cattle or horses at it.

M. Bennett]   ARMOURED TRAIN.   [*Photo.*

If suddenly confronted with such a fence when under fire, it is best to *lie down*, and contemplate the best method of getting over. The wires and posts may even be cut while so lying.

Wire netting, over 3 feet high, and with one or two strands of barbed wire running through it, makes a good unsurmountable obstacle, taking a long time to cut through.

Abattis was, owing to the dearth of trees and firewood, not much used in South Africa. A belt of it was placed round Mafeking, and at Warmbaths and Pienaars River and other northern places it was used to advantage.

Portable fences were occasionally carried with a force on the march to erect between or in front of outposts. The wires may be permanently fixed to the posts, which, if two-legged, may simply be placed on the ground, and each end of the fence securely pegged down.

As regards position of obstacles, it is better that they should, if possible, not be in the line of ordinary fire from either side, else they may be cut by bullets; yet they should be well under fire from the position.

If material is plentiful, wire fences are best arranged so as to run diagonally away from a work, instead of directly across its front. The enemy, in charging up, is then very likely to run alongside the fence, in preference to getting over; and a number of men may thus be enfiladed, or caused to crowd together and form a good target.

Obstacles should be close enough in front of a work to be within range of view and hearing at night or in a fog. If really impassable obstacles, they may be close up; but if the object is only to give warning of attack, they should be fully 100 yards off.

Obstructions. Obstacles and obstructions may also be useful, outside the range of firearms, in delaying the advance or retreat of an enemy over a certain route. Thus *roads* (especially in passes and defiles) may be blocked by felled trees, by blowing an excavation with dynamite, or, temporarily, by lighting a large fire on them. *Bridges* may be destroyed, or blocked by the same means. *Railways* rendered useless by blowing up or removing rails, or even temporarily by burying under a heap of earth or stones, which can be rapidly effected by a lot of men even without tools. A frequent practice of the Boers was to place small charges of dynamite under the junction of rails which, when exploded, caused each end to turn up. This method, while using a minimum amount of explosive, necessitated much labour to repair, as two entire new rails were required.

Roads may be temporarily blocked against a rush at night, by collecting a tangled mass of wire (a frequent result of a destroyed fence), which can easily be dragged into place and on to one side as required.

At Zillikats Nek, a felled tree, with its branches intact, was dragged into place on the road, and removed, when required, by a team of mules.

Fords and fordable rivers may be rendered practically impassable by deepening with trenches or dams, or by placing obstacles, such as barbed wires or abattis, under water.

To prevent guns and waggons crossing over strategic lines in the Transvaal, *ditches* were dug between block-houses, over which it would be impossible to get wheeled vehicles without delay, which would enable the troops in the block-houses to open fire upon them.

CHAPTER VI.

OUTPOSTS AND PROTECTIVE SCREENS.

PROTECTIVE guards—called outposts when a force is halted; advanced, rear, and flanking guards when a force is on the move—have always been recognized as institutions of very great importance to an army, and the recent campaign has exemplified, most clearly, how much depends on this duty being thoroughly well executed.

It is true that in a perfectly open country like South Africa, the arrangement of outposts is more difficult than it would be in districts intersected with hedges and ditches and walls, and in which well-defined roads are the only means of advance for an enemy in force. And as regards protection while on the march, the absence of impediments made it comparatively easy.

Most of the reverses we sustained in the war were due to neglect of some sort in providing a proper protective screen. Sometimes it was that the outposts were badly placed round camp. Sometimes the sentries were not on the look out. Sometimes the advanced guard was deficient.

Sometimes the rear guard was at fault. But, whichever was to blame, the main body, deprived of the full use of its feelers, was taken by surprise, and came into contact with the enemy without being fully prepared.

These duties are closely connected with, and dependent on, reconnaissance. If the mounted troops carry out their duties as they should, a commander should have full information of the presence of any large body of the enemy within at least 10 miles of him, and should even be able to get a good idea of whether an attack on him is likely. And if the mounted troops are active and enterprising during the day, there is much less likelihood of the outposts being disturbed at night.

With the Boers, owing to their mobility, and what we may call their "dispersive" tactics, it was seldom possible to keep in touch.

Outposts.

Outpost duty may be carried out in a variety of ways, and many different systems were used in South Africa. While some commanders and some corps were very particular about its being properly conducted, others were undoubtedly slack in maintaining its efficiency. The Boers generally got to know of such details, and always avoided attacking those who proved themselves vigilant, cautious, and resourceful in their methods of protection.

The subject, then, is one that should receive

202 OUTPOSTS AND PROTECTIVE SCREENS.

more careful attention. The very objects of outposts seem sometimes to be forgotten.

Objects. The primary object may be considered as being to *check an attack* until the main force has had time to prepare for it. Secondary objects are to prevent snipers, and even guns, from taking up a position whence they can bring fire to bear on the camp, and also to prevent hostile scouts from obtaining information. But the comparative importance of them will vary with circumstances, and will involve different dispositions. Much will depend upon whether the outposts are placed around a force on the march, to give it security while halted during the night, or if to protect a camp likely to remain for some time in one spot. In the first instance, as the troops will generally be getting ready to move soon after (if not before) daybreak, artillery and long-range fire need hardly be taken into account, nor will it probably be important to keep hostile scouts at a distance, and the main object will be to hold in check a rush of the enemy. As, moreover, the site of the camp will be chosen rather to suit the length of the day's march, and as providing good water and other supplies, but little choice of ground for tactical requirements will probably be allowed, and therefore greater strength will be needed.

In the second case, good positions would usually be carefully chosen and artificially improved, and posts can consequently with safety be pushed further out and be more weakly held.

So a large force may have no fear of being actively attacked, yet may wish to guard against disturbance from occasional snipers; or it may be of importance to keep its strength or composition unknown to the enemy. So that, in such a case, it will be necessary to arrange the outposts a long way out.

It is sometimes considered that to *give warning* of the approach of an enemy is the chief duty of outposts. *If* such warning can always be given with certainty and in plenty of time, it may suffice; but practice has shown that attacks by night can be made with such suddenness that, if the outposts are not capable of offering resistance, they must be a very long way out and able to immediately warn the main body, else the latter will not have sufficient time to form up.

At Belmont the Boers had a small post a mile or so in front of their position. On our approach in the early dawn, instead of firing, they ran back, but were only able to give warning when we were close on to the main position.

Detached posts a long way out are always liable to be avoided or cut off by those advancing to the attack, while if so numerous and close together as to obviate such a contingency, they would absorb a great number of troops, and are then able to offer resistance.

To be capable of offering resistance is, then, the main point.

According to the accounts published of the attack

on Colonel Firman's camp on Christmas Day, 1901, the Boers approached the position during the night, and as soon as challenged by the sentry, they "rushed into the British camp, shooting our men down point-blank as they came out of their tents," which seems to imply that the outposts could not have been disposed to stay a rush, and that they must have been so close in as not to be able to give timely warning.

To at once alarm the actual outposts when attack is imminent is, of course, of the utmost importance. The ability to delay the advance of an assailant must depend chiefly on the strength which can be opposed to it; but as the enemy will almost certainly be much more numerous than the detachments barring their advance, man to man fighting at close quarters must be avoided, and it becomes urgently necessary to have ample notice of an assailant's approach, so as to open a rapid and unexpected fire. Such opportunity for preparation must be dependent on the alertness and on the judicious posting of sentries.

Strength. The strength of outposts, though often hitherto considered as necessitating the detachment of a certain *proportion* of the force, is really solely dependent on two points: 1st, The extent of front to be watched and guarded; 2nd, The suitability of the ground to defence and observation.

The first, if it comprises a circle around a camp, will not vary so very greatly with the size of the main body, and the strength, varying according to

OUTPOSTS AND PROTECTIVE SCREENS. 205

whether attack is actually expected, or there is a certainty of no force of the enemy being within striking distance, will usually be a given number of men per mile. Under ordinary circumstances 100 men per mile should suffice to temporarily hold in check an attack.

In order to bring the strongest opposition to bear at any point, it might seem desirable to have the force on outpost concentrated together, ready to be moved to any threatened point; but, since each man should, as far as possible, be ready *in situ* for immediate action, the force must be chiefly distributed along the outer line of defence. The first consideration is, then, what distance out from the camp this line should be.

It is sometimes urged that outposts should be pushed well out from camp, so that, should they be heavily attacked and driven in, the main body may have plenty of time to prepare. Yet it must be borne in mind that the further out they are placed, the longer the line of perimeter they will have to occupy, and therefore the weaker they will be. With outposts close around the camp, though the length of front may be smaller and consequently able to be more strongly held, there is the objection that, if driven in, they are so close that no time is allowed the main body to prepare (though, if they are thus comparatively thickly posted, they should not have to retire so precipitately). Also that if a hot fire is directed on them, bullets will pass into camp, so that one of the secondary

<!-- margin note: Extent of front. -->

objects of outposts is not fulfilled if they be too close in.

With a very small force (under 1000 men), the only way to be really secure is to adopt a different system, and have the main outposts close around the camp, and all must be liable to suffer from the fire poured in.

When we come to consider protection from long-range rifle fire, we have a definite basis to go upon. If the enemy's weapons are capable of doing effective work at 2000 yards, it stands to reason that if the main body is to be safe, the outposts should be nearly that distance from camp (presuming the country to be quite open). This means that, if the camp be fairly compact, a circle of some 15,000 yards round must be taken up. Such an extent of front may be very well for a large force, or for one well intrenched, but, as a rule, it will be too great for a small independent force to temporarily occupy in sufficient strength.

If it is desired to protect the main body from disturbance even by shell fire, a very much larger area will have to be guarded. Should the enemy possess heavy guns, able to throw projectiles up to 6 miles, it becomes practically impossible, for any except a very large force, to provide against, although natural features of country would usually intervene, and be taken advantage of.

Modder River may be taken as an instance of a large standing camp which it was desirable to protect against artillery fire. Detached works were

OUTPOSTS AND PROTECTIVE SCREENS.

erected all round, perhaps 2 miles from camp. The Boers, however, were not able to attain any good position from which they could see to shell the main body, and contented themselves with occasionally sending a projectile hurtling through the air over the hills in the direction of the camp, but which invariably fell short on the open ground behind the outposts.

But characteristics of ground and other circumstances will greatly modify the extent of the line necessary to be occupied. If a force be encamped in a favourable situation, such as a hollow closely surrounded by a ridge of low hills, a smaller circle may be held than if on an open plain, since the hills may be occupied by the outposts, and then snipers are not likely to be able to fire into camp; scouts cannot see the main body, and, if attacked, the outposts, being comparatively close to the main body, may soon be reinforced. But if the position be in the flat, or with hills commanding a view of it, matters become more difficult, and the outposts, as a rule, will have to be pushed much further afield.

It would, however, be very rarely the case that the country would be so open all round as to necessitate pushing out the outposts to the distance above referred to. Rising ground will generally exist to one side at least, giving protection from aimed fire. Neighbouring heights and advantageous spots may be held by detached posts, if well supported by the fire of the main outposts, and by this means even the enemy's guns may be practically prevented from doing much damage. Very often, too, when the country has been thoroughly well scouted, the advance of the enemy may only be expected from one direction; so that, though it is always desirable to be prepared against all contingencies, the rear side may be but weakly held.

Taking an ordinary instance of outposts surrounding an independent force of a few thousand men, about 1000 yards would usually constitute a suitable average distance for the main line of defence to be posted from the main body. Allowing an area of about 700 yards across for the camp, this would imply that the extent of front to be guarded would comprise a circle of about 8000 yards, or 4 to 5 miles, in circumference.

Suitable natural features of ground may allow of this length being slightly reduced, and with a large spreading camp or unfavourable environments, a still greater line may have to be occupied.

It may, then, be taken that, allowing 100 men per mile, about 400 to 500 men (or say half a battalion) should usually suffice to occupy the main line of outposts for a force of moderate size—that is to say, one that can conveniently camp in a space 700 yards across. Reserves may be added according to circumstances.

Dispositions. As has already been said, the systems of disposing the troops on outposts during the war varied considerably, but they were not usually in accordance with the regulation methods previously

OUTPOSTS AND PROTECTIVE SCREENS.

described in drill-books. Outposts were seldom, if ever, divided, as there laid down, into piquets providing a number of sentry posts, supports, and reserves, but consisted, as a rule, of a chain of posts with one sentry over each, forming a line of defence. Posting a considerable number of men in the front line, ready to open fire at a moment's notice, especially when good positions can be chosen, and artificially strengthened, is decidedly preferable to placing them as supports somewhere in rear, and having to be moved up when firing begins.

The reasons for these changes are not far to seek. With the rapid discharge and far-travelling bullets of modern musketry, movements under fire are dangerous, and it is preferable for all troops to remain where they are posted. Long-range fire enables a greater extent of ground to be covered, and the exact relative position of the defenders therefore not so essential, so that falling back and reinforcing should be avoided as far as possible. But it is as regards the strength of these groups or piquets, and their intervals apart, that considerable diversity of opinion was shown. Mounted corps sometimes relied solely on a few cossack posts in favourable spots, often miles from camp, to gallop back and give warning of the enemy's approach. In small mixed columns one side was sometimes protected by perhaps half a dozen infantry piquets of about 20 to 30 men, while on the other side the cavalry threw out only 3 or 4 posts of 6 men each

210 OUTPOSTS AND PROTECTIVE SCREENS.

to guard an equal amount of front. The system of a few small posts far out is certainly not to be depended upon, at all events if the troops in camp are not so disposed as to be ready to immediately repel an attack. If too far apart, the enemy may easily slip in between them. Should the sentry of one post not be alert, the post may be quietly captured, and a wide opening allowed to the attackers. Being so far away, such outposts may not even be able to alarm the camp, especially if a strong wind be blowing towards them.

A cavalry force near Machadodorp detached a small post to occupy an important hilltop. This was surrounded and cut off by the enemy, and owing to the distance off, and to a strong wind blowing, the firing was not heard in camp, and the main force was taken by surprise.

Small posts, far away, are liable to panics. A few men by themselves have often been known to get "jumpy," and come in, declaring the enemy to be advancing.

Each piquet most usually consisted of a section, and these were placed at about 400 yards apart. If the ground between them was not well seen, or the piquets considered too far, a group, consisting of a non-commissioned officer and six men, was posted between.

A group of this size was always found preferable to the old-established rule of three men to a sentry-post; which undoubtedly makes the duty come very hard on men who have been

OUTPOSTS AND PROTECTIVE SCREENS. 211

marching all day, and may have to continue marching next day. With six men, each can get five hours' continuous sleep, even if relieved hourly. The more men there are to a sentry-post the less irksome is the duty. But it also allows *double* sentries to be temporarily mounted when necessary, such as to ensure vigilance when attack likely, to better watch difficult ground, or if a very dark night. These sentries were usually posted well out on each side of the piquet, and thus covered more ground.

An occasional big post is desirable for several reasons. There will often be some one road or approach along which an enemy is more likely to come than elsewhere. Indeed, in cultivated country night advances would be confined to the roads. This may be more strongly guarded. Often after the outposts are in place, it is found desirable to add another post where the line may be weak, or special spot requires guarding. The men for this can be taken from the strong post.

The distance apart of piquets, or rather sentries, is limited by the proximity within which an enemy can creep. If posts are as far as 600 yards apart, an enterprising enemy, having ascertained their exact whereabouts, could, on a dark night, wend its way cautiously between them. Two hundred yards is about the furthest that a sentry can watch with certainty, so that 400 yards is about the greatest interval that should exist between posts. On a stormy night, or if the ground be covered with bush, they should be closer. It is also desirable to have sentries fairly close, since the safety of the whole force should not be entirely dependent on the vigilance of one sentry. If one should fall asleep, the sentries on either side should be able to give the necessary alarm of an approaching enemy.

On a bright, still, moonlight night, a sentry can often see and hear almost as well as by day; but on a pitch dark, stormy night, or a thick, foggy morning, it is very different. Yet by regulation and practice no different dispositions are usually made.

And it must be remembered that even on a fine night, clouds may gather over, strong winds blow, or a thick fog come on, so that it is not prudent to put the sentries too far apart. Perhaps the best method of meeting this difficulty is (as already mentioned) for extra sentries to be temporarily mounted whenever considered necessary by the officer in command of a piquet.

Another system of outposts sometimes used, especially if a night attack seemed imminent, was to closely surround the camp with a complete cordon of posts, not more than 100 to 200 yards apart, and about that distance away from the actual camp. These posts, with some sentries well in front, formed a practically impenetrable barrier to a rush, while away outside this, 500 or 600 yards off, were a few posts in well-chosen positions, and intrenched, not only against attack from

OUTPOSTS AND PROTECTIVE SCREENS.

outside, but with a bank or other arrangement to protect the men from the fire of those behind. These, if judiciously posted, would generally be able to give warning of an attack, and would greatly disconcert the enemy, while, being so intrenched, need not mask the fire.

Though this arrangement should almost infallibly prevent an attack from succeeding, it does not prevent damage being done among the men and animals and even vehicles in camp; but it is practically the only safe way of protecting a very small independent force.

Marching from Springfontein with a small force (two companies infantry, two guns, and a number of waggons) I had information that a night attack on us was contemplated. We halted near a small fortified post on the railway, the guns and waggons were placed close to this, and the infantry were then disposed in a sort of semicircle around, each man lying with his rifle beside him, just as if taking up a line of defence by day. In order to cover the necessary extent of ground, large gaps had to be left in the line, and outposts, which consisted of small groups, were placed about 150 yards in front of these gaps. Thus, in case of attack, all that was necessary was for each man to fire his rifle to the front. Practically no time was needed to prepare for defence. Though the enemy hung around all night, blew up the railway within 800 yards of us, and several shots were exchanged, they did not press the attack. It would have been practically impossible for so small a force to guard against snipers and shots coming into camp; or against scouts obtaining information, as the whole party could easily have been counted to a man from a neighbouring height.

The Boers sometimes adopted a form of outpost which may be called a "buffer post," consisting of a small mobile force, perhaps twenty to a hundred strong, all mounted and free of impedimenta, posted some miles from the main force in the direction in which the enemy is likely to come. This party provides its own outposts, and if attacked, whether by day or night, while making some show of a stand, will at once send back to warn the main body, who can then make full preparations for defence, or, as more frequently happened with the Boers, start off their convoys and beat a retreat. With intelligence of the enemy's whereabouts this would usually form a valuable addition to the outposts.

Posting outposts. In deciding on a camping ground, one of the most important considerations is suitable positions for outposts. This is an item sometimes apparently scarcely taken into account by staff officers, and the outposts being badly placed, they are surprised if the piquets are driven in.

If the occupation be for several days, it will be usual to carefully select positions and intrench them. But if merely halting for the night, the positions taken up will not be so important, and the length of march, water supply, and strategical

OUTPOSTS AND PROTECTIVE SCREENS. 215

considerations will, perhaps, be more considered than the tactical suitability of the ground for the outposts line.

There is always a great difference between the disposition of outposts required *by day*, and those *by night*. In the former with open country and elevated positions an efficient look-out can be kept by a very few sentries, and, as the enemy may be seen from afar, longer notice can usually be given for reinforcements from camp to be brought up, and positions, quite apart from those occupied by the sentries, chosen for defence.

It was a very usual practice when a chain of piquets had been placed for night work, to withdraw perhaps two out of every three during the day.

As a rule the strength of outposts in daytime will entirely depend on the nature of the country. If it be very open, though a single sentry may be able to give sufficient warning, yet we must always be provided against the possibility of one man not being vigilant; moreover, thick weather may come on and interrupt the view.

But at night the men must be in position, ready to offer resistance against a sudden rush, so as to hold the enemy until the troops in camp can turn out. So long as the ground is open for a few hundred yards around and some cover is available for the piquet, nothing more is wanted. It will be practically impossible to prevent hostile scouts and snipers from hanging around; a view extending

Position by night.

far to the front does little to assist in this matter, and if it only extends as far as a good close sky-line it may be preferable to a view extending for miles; which latter is only more likely to give away the position of the outposts to the enemy.

Hilltops are therefore not specially valuable if the posts are to be evacuated in the morning, except that the enemy must be prevented from occupying them. Several posts round the base of a hill will generally be better than if placed on top, while valleys and river beds have to be specially watched.

Some officers have a strong objection to posting their men down in a hollow, but although the idea may sound rather like giving the advantage of ground to the adversary, yet it is well known that such a position really possesses many advantages as a night post; notably, forcing the enemy to show against the sky-line, while the post itself lies in the shade. The line of fire is more grazing and not plunging, and, as a rule, the position of the piquet is not discernible from a distance.

This is just an instance of where command of ground confers practically no benefit on its possessor, since no distant view is possible at night, cover for assailants not usually important, guns not usable, and even covered retreat seldom necessary.

Care must always be taken at night by posting extra sentries, or pushing out groups, not to allow a few men to slip in between posts, and fire on them from behind. At Nooitgedacht and other places the Boers succeeded in doing this.

If commanding positions occur in the neighbourhood and there is a probability of the enemy placing guns on them, it is usual to occupy them with mounted cossack posts. If the latter be driven off, they will at all events give notice of the enemy's presence, and dispositions can then be made accordingly. It may often be undesirable even to risk leaving such posts away by themselves at night; in such a case they may be sent out before daybreak, and will then be able at all events to ascertain if the enemy is there.

Generals and staff officers seem apt to forget the difficulty there is in posting outposts *after dark* on unknown ground. This is really an important point to bear in mind. It is impossible to select good positions under such circumstances, and if an attack should come, the outposts may be taken at a great disadvantage. Not only are natural features of the ground not fully taken advantage of, but the various components of the outposts are apt to get mixed up. I have known such a case where, after a piquet had been in position for some time, it was discovered that a piquet from another regiment had been posted a few hundred yards directly in front of it.

After the battle of Magersfontein, when there was undoubtedly some excuse for confusion, there was considerable difficulty in arranging the outposts. Men of various regiments seemed to come up in driblets all through the night, and were made up into piquets and plumped down wherever there appeared to be a clear space.

One of the best methods of posting outposts is for an officer to select the positions by daylight. The troops can then be marched out after dusk (the cavalry meanwhile being responsible for the safety of the camp), and this generally allows them more time and rest after their march, and prevents watchers of the enemy noting from afar the positions of the outposts. One or two officers riding about at their leisure can much better select positions than if the troops are on the spot awaiting definite orders. If the force on the march is not likely to reach its destination before dark, an officer can usually be sent forward to select positions for the outposts.

Sentries. The duties of sentries being to see without being seen, the system in vogue in peace time of pacing up and down is bad. If only men could be depended upon not to go to sleep, lying down would, of course, be the best position, but unfortunately this tendency cannot be avoided, and to stand erect is the only attitude which can be allowed. Though sentries caught in the act are often apt to plead they did not think "there was any harm" in sitting down, and though in some of the irregular corps it almost amounted to a practice, yet it must, of course, be put down with a firm hand. By sitting, and especially by lying down, a sentry may rest himself, keep well out of sight of an approaching enemy, and be in better position for opening fire. Yet it is very certain, as all will agree who have tried it, that the mere act of sitting, or even leaning against a support, is very conducive to sleep, and for a tired man to lie down is almost certain to prove fatal to vigilance.

With such instances before us as the night capture of the entire Yeomanry camp at Tweefontein, the supreme importance of alertness on the part of sentries is seen to outweigh all other considerations.

If they stand perfectly still, and especially if there be bushes or rocks about, sentries will not be easily seen. But here, again, another practical difficulty crops up. Nights will be cold, and to keep themselves warm sentries must walk about. Often, too, it is not easy to get a complete view from one spot, and they must occasionally move round to see that all is quiet.

In order to prevent a sentry being conspicuous on open ground, a small pit about 3 to 4 feet deep was occasionally dug, in which he stood.

A system sometimes advocated is that of double sentries on the supposition that one will keep the other awake. I know of one instance, at least, where *both* sentries of a double post were found asleep.

Another custom of peace time which is objectionable on service is that of a sentry challenging, in a loud tone of voice, all persons approaching his post. As often as not he knows perfectly well who it is that approaches, so that his noisy challenge, betraying his position to any scout a mile off, is worse than useless. It would be better only to challenge when really in doubt, or indeed, some officers instructed their sentries never to challenge a single or even two or three persons approaching. If an enemy, let him come on and be captured, or shot at close range, instead of frightening him away at at a distance. The usual form of challenge, too, is very unpractical. A Boer at all events would always answer "Friend" to the challenge, and steadily advance while the sentry is going through his long recitation. "Halt! advance one" is quite enough when he sees an unknown party approaching.

A countersign was very generally omitted on the march. If an enemy is coming on to attack, the request for the countersign will not delay matters. If a hostile spy approaches, it will be very probable that he has been able to ascertain the countersign. I remember, in manœuvres once, a reconnoitring patrol, wishing to ascertain the enemy's countersign, waited for one of his patrols and challenged it, demanding the word, which was of course promptly given. In an enemy's ranks there will often be some one sufficiently conversant with our ways and customs to try such tricks.

"When in doubt, fire," is a good rule for sentries. Though the act of shooting a cow or a stray donkey in mistake for an attacking foe may appear ridiculous, it does but little harm, and is far preferable to making the fatal error of not giving immediate alarm when an enemy is really advancing.

OUTPOSTS AND PROTECTIVE SCREENS. 221

The question of supplying fully efficient outposts is one of the great problems of war. Watch-dogs, which have been used in a few instances, would appear to be a most valuable aid, and perhaps in a future war we shall see a properly organized body of well-trained dogs. Spring guns and alarm wires might also be more used to great advantage. Portable obstacles, such as barbed-wire fences, are most useful for preventing posts being rushed. *Other details.*

Another important alteration in the old system of outposts which was frequently made was the abolition of regular patrols. Though in some corps patrolling was continued, especially in guarding the railway lines, the arguments against the system, and the paucity of opinions in favour of it, caused it to be generally abandoned. Much will, of course, depend on the circumstances prevailing, but it is worth carefully considering the *pros* and *cons*. *Patrols.*

Patrols are supposed to fulfil the following duties:—

1. To ascertain that the sentries are alert. But even a sleepy, reclining sentry will usually rouse himself on hearing the steady tramp of the patrol, and so soon as it is past he knows he is safe from disturbance for a long time. With the now prevailing system of each piquet finding one sentry, visiting patrols are not necessary, since the non-commissioned officer in charge can see the sentry.

2. To make sure that the enemy is not encroaching between the piquets. But if a few of them were stealthily trying to slip in, the chances are

that they would see or hear the patrol before the latter would see them, and would then lie low till danger was past. It is not as if all our men were born scouts, able to creep about noiselessly while keenly watching for every sign of an enemy. The average soldier is dull and careless, and performs his duty in the easiest way to himself. Moreover, he will not be expecting the enemy at any particular moment. On the other hand, the enemy's scouts or guides will be specially picked men eagerly on the alert as they approach the vicinity of the outposts. If it be known that patrols do go round regularly, these would be watched for, and no advance made by the attackers till they were past. Again, if the outpost groups be not more than 200 to 400 yards apart, it should be practically impossible for any one to slip through.

3. To move out to the front to see if the enemy be advancing. But here, again, the enemy might watch for them, and await their return before advancing, or arrange to cut them off.

The actual objections to patrols are—

1. Liable to lose their way in the dark. Then not only are they useless, but they may disturb others. Those who have done much night duty will understand this difficulty of finding the way. It may be all very well in certain places when good landmarks, such as trees or houses exist, but on the open veldt there were, at first, innumerable cases of patrols being absolutely lost all night, even though within a few hundred yards of their piquets.

2. If enemy's scouts be watching visiting patrols, they can ascertain the exact line of front and position of each sentry, especially when the latter challenge.

3. The tramp of marching, especially among rocky kopjes or on the ballast of a railway line, can be heard from far off on a still night, so that the enemy can ascertain the position of the outposts and know when to avoid the patrols.

4. Even though they go noiselessly, if on high ground, as is usually the case, the patrols can be seen against the sky-line.

5. If the patrols only go round at long intervals, they form no guarantee of safety.

6. If men have been marching all day, and have to start off again the next day, it becomes a very irksome duty to march several miles during the night.

There may be cases in which reconnoitring patrols will be useful. In thick bush, where no distant view is obtainable, or, indeed, in all enclosed country, they may seem more suitable than they did on the open veldt. Yet they will seldom do much good, and would be better formed into extra sentry-posts, ready to receive an enemy in their well-selected positions. If the view be very restricted, it may be expedient to send out some form of patrol when an attack is actually expected, or to make sure that all is clear before outposts withdraw, if they are to do so before daylight.

A patrol may also be sent out when outposts first mount, to ascertain the lie of the land and position of the neighbouring posts. A single picked scout will usually do even better than a patrol. He can see as much, and will be less likely to be seen or heard.

If it be desirable to watch special points, or give better warning of an expected attack, the patrol should move out to a suitable spot, and then lie down and watch. This system may be useful, too, in a place long occupied, where the enemy may have ascertained the exact positions of the regular piquets, and may try to slip in between.

If patrols are to move at night, it was found desirable and efficient to place white stones, whitewash patches, or (temporarily) even torn-up paper, at intervals along the path.

As a substitute for visiting patrols, it was a usual custom for officers or non-commissioned officers to take their turn "on watch," during which they were to keep awake and keep an eye on the sentries, often of two or three groups. Sometimes they would stroll quietly round the posts, generally well in rear of the line of front, so as not to be seen by an enemy, and avoided being challenged.

In small permanent posts, when the commander occupied a central position with sentries around, some simple method was sometimes improvised, such as the pulling of a wire, by means of which the officer could ascertain if the sentries were alert, they answering by another pull. Even such a system as for the sentry to strike a match (screened from the front) in reply to a lamp exposed, might answer the purpose.

Fires at night, in open country, of course give away the position of outposts. But they are often very essential for cooking and for warmth, and if lit well away to the flank or rear, the enemy may only be confused thereby. I was once, in a former campaign, with a small detached force, when we lit a number of fires all round our position at night which would, we hoped, not only tend to give the idea of the existence of a large force, but would also enable us to see the enemy if he advanced to attack.

Fires.

At Colesberg the line of day outposts was shifted back every night, but fires were kept burning to give the idea that the same line was held. A Basuto with a supply of fuel held the outpost line against the Boers for a fortnight.

Outposts are very usually ordered to stand to arms for an hour before dawn. But except under special circumstances this seems hardly necessary, rendering the duty more arduous and exhausting. If all the sentry-posts be doubled during this period, it will usually be sufficient to ensure alertness. To make sure of the men being awake, they were often ordered to stand up in their places, and one then saw on the rising undulations of the veldt, each group standing like a row of ninepins against the dawning sky-line—a most instructive sight for the possibly advancing enemy.

Standing to arms.

226 OUTPOSTS AND PROTECTIVE SCREENS.

Outposts should not be withdrawn too soon. A small column camped near Cradock, and, knowing the enemy was near, put out strong outposts on the surrounding hills. After daylight, just as the column was preparing to move off, the outposts were drawn in, but their positions were almost immediately occupied by the enemy, who thus surrounded the force and kept it in an awkward position all day.

Supports and reserves. Supports or connecting posts may be necessary in certain particular cases—such, for instance, as a mounted post some distance from the main body. If driven in, a support posted half way will enable it to retreat more leisurely. It will also be able to pass back word, and warn the camp.

Though very seldom employed in the late war, some form of reserves may often be necessary for outposts, especially when a large force has to be protected. The chief idea would then probably be to have a force in one or more central positions, ready to move at a moment's notice to any one of some previously selected positions. Or it may be preferable to have the force ready distributed in these positions.

Though regular outpost duty involves strain and lack of rest to those performing it, to sleep with a detachment away from camp, to do without tents, and to sleep with boots on, is no such great hardship. So that reserve duty is not exhausting to troops.

Occasionally, certain troops in camp were told

OUTPOSTS AND PROTECTIVE SCREENS. 227

off to support or reinforce particular piquets in case of alarm, though a better system is to arrange that all men in camp can be quickly made available to repel attack.

A few guns may well be told off to be in reserve ready to move out in case of attack. It will seldom be worth while for them to be among the advanced outposts.

As regards the reinforcement of outposts, we must again remember that it is, as a rule, not more *men* required at a given point, but more *fire* brought to bear in front of it. It will, therefore, not usually be desirable to push men up to the particular trench or position held by the piquet, which would only give a better target to the enemy, but to move them up to some advantageous spot, perhaps halfway between two posts, from which they can bring a telling fire on the attackers.

It should be the duty of the officer posting the outposts to note desirable positions for the reinforcements to occupy, and commanders of posts should be acquainted of them so as to be able to direct the supporting troops.

Another important component part of outposts are the sentries and guards in camp. On them devolves the important duty of arousing the main body when the outposts are attacked. It seems highly probable that in some of the night assaults in which we have severely suffered, the men in camp have not been awakened in time. Often, *Inlying sentries and troops in camp.*

too, the troops in camp were not so disposed as to be able to fall in at a moment's notice on an interior line.

Should there be any likelihood of an attack, which is, indeed, a contingency always to be guarded against on active service, men would, of course, be ordered to sleep with their boots on, and magazines of their rifles charged (bandoliers, etc., are often mislaid in the dark). We have had several instances of men, on a sudden alarm, being quite unprepared (*vide* the instances quoted on pp. 204 and 230). In one case, the arms were piled, and equipment hung on the piles, which caused serious delay.

There are even instances of outposts in some standing camps being provided with bell-tents—a most palpable mistake, as, not only do the tents show up the position of each piquet from miles off, and form a good target for the enemy's musketry, but also bottle up the men and cause great delay in forming up. Those who have had to go without tents for months on end in cold and wet, will know what a mere luxury they are. A small, low-lying shelter of blankets can be made nearly as comfortable.

When a force is camped in a locality not very suitable for defence, a position should be selected and pointed out beforehand, to which the force would be moved immediately it became evident that the enemy were advancing to attack.

The following account from the *Daily Telegraph*

of the affair near Bethel on June 12, 1901, though only professing to be second-hand evidence, nevertheless suggests many useful hints on outposts :—

"General Beatson, to whose mobile column they were attached, when near Bethel, sent the Victorians, about 260 strong, with two former Boer pom-poms, all under Major Morris, R.A., ahead to scour the neighbourhood, and bring in some cattle. When twelve or fifteen miles from Beatson's camp, they got in touch with the Boers, who were found to be in very considerable strength. As a matter of fact, the enemy was under Ben Viljoen, who had with him a Commandant Muller—the same pair that captured the 4.7 gun "Lady Roberts" at Helvetia. After skirmishing, the V.M.R. formed camp quite early in the day (the 12th), expecting to have a fight next day, but not that night. Major Morris gave orders for placing the outposts, which were not numerous. Unluckily *a valley, or depression, was left unwatched*, although men were put upon the crests of the ridges. It was *up that dip the Boers stole* to deliver their attack. Another evil was, that the posts were taken up in daylight, and not changed when darkness fell, so that the enemy knew how to avoid them. It deserves noting that *at night men placed in a valley* can much more easily detect any movement, as they have the skyline for a background, than a picket upon a hill. The Victorians turned in early on the 12th inst. Indeed, what else can troops do upon the bare veldt when night falls, and with it descend chill hoar frosts ? For then you are snugger if rolled in your blanket and waterproof. For the most part the arms *were piled, an unwise adherence* to drill regulations, now more generally 'honoured in the breach than in the observance.' Brer Boer knows better, and by day carries his rifle slung over his shoulder, whether afoot or on horseback, and he sleeps with it lying handy by his side. It is his weapon, his particular property and charge, and not to be stacked and shuffled with promiscuous rifles. No sooner had darkness set in than the Boers, who were but a few miles off, advanced to surprise the Victorians. *Guided by traitors* they came on up the valley, passing the piquet without challenge. Viljoen posted part of his men to command one side in case of accident, and Muller, with about 180, stole towards the camp. All was quiet, save the sound of the horses, which were securely picketed in a double line upon the east side, munching the corn and mealies the Australians had spread out for them. The saddles and bridles were ranged behind. By them most of the troops slept, but there were soldiers lying about by the carts, baggage, and guns. In the first—and last —sleep of many about 8.20 p.m. there burst over the still camp a wild roar of Boer musketry, fired from a range of fifty yards. Half awake and dazed, soldiers scrambled out of their wrappings and ran to grab their rifles from the piles. But the Boers covered the stacks and shot the troopers down. The officer in command of the pom-poms

BOER GUN-EMPLACEMENT.

ran to cast one of them loose and use it. He was riddled with bullets in an instant. And still the shouting and slaying went forward, the enemy rushing wildly about the camp."

It seems undoubtedly probable that the outposts on this occasion were most faulty, although it is evident that there were traitors in camp—a contingency that should always be allowed for—who were able to guide the enemy to the weak spots. Had the men been lying round the edge of the camp, ready to shoot outwards, it is doubtful if the Boers could have got inside. In the end, the camp with all the guns and stores was captured, but very few men escaping, and our casualties amounting to about 60, though the Boers also lost heavily.

Advanced, Rear, and Flanking Guards.

Advanced, rear, and flanking guards are but the equivalent of outposts on the move. But there are these differences. Even in daytime, and in comparatively open country, the scouts of an advanced guard are liable to come suddenly on an enemy, which is the equivalent of an enemy suddenly attacking at close quarters; and secondly, an advanced guard is unable to occupy at any moment a favourable position and be ready to repel attack.

A force on the move is always much more vulnerable than one occupying a chosen position,

232 OUTPOSTS AND PROTECTIVE SCREENS.

so that an efficient protective screen is of even more importance on the march than in camp.

As these guards have to march along at the same rate as the column they protect, and have to do the fighting as well, great mobility is essential.

The general principles of advanced and rear-guard action are the same as ever, but two innovations have to be taken into account in carrying out the details. Firstly, increased ranges necessitate increased distances between the various component parts; and secondly, the improved power of defence permits of comparatively smaller bodies being used for the purpose.

But these facts are often not taken into account. In a recent article on the war by a well-known writer, General Yule's retreat from Dundee is compared with that of Sir John Moore on Corunna, and the warding off of the pursuing forces is treated of as though the cases were exactly similar. But think of the difference! If Sir John Moore had drawn up 1000 of his infantry to check the pursuit, these men would have had to wait till the French were within a range of not more than 200 or 300 yards before they could do anything to delay them. Then, if the French pressed on in greatly superior numbers, it seems hardly conceivable that the British could have made further stand; they would be bound to suffer serious loss, and would have had to retire with the enemy on their very heels. Even if cavalry had charged the head of the pursuing forces, they would have to gallop back to catch up the column, and would then very soon become worn out.

In Natal, on the contrary, if only 100 men lined a good position and opened fire at 2000 yards on a pursuing enemy of ten times their number, they could force the latter to deploy for attack, and to advance slowly and cautiously for nearly a mile, when the little force could retire with impunity to a fresh position and repeat its tactics indefinitely, causing the pursuit to be both slow and harmless.

The chief duty of advanced guards is to reconnoitre. They are usually composed of cavalry and horse artillery only, as these are sufficient for the scouting duties, and if they became engaged, could hold their own till infantry came up, or could rapidly retreat if necessary. *Advanced guard.*

Owing to the great range of guns, it would not be advisable to move infantry in close order with an advanced guard, and to move them extended unnecessarily fatigues the men. But some infantry, preferably in a more or less extended formation, must move ahead of the column ready to support the cavalry if driven in.

Advanced scouts, widely extended, lead the way, with supports perhaps a mile behind, so as to be practically out of range if the scouts become ambushed.

The ground bordering the route, not only to the direct front, but also to the flanks, must be reconnoitred. As the enemy may be encountered at any moment, and as one man can see as much as a dozen, it will be safest and most economical to have very few men out in front. Their number will depend entirely on the country. If bushy and broken a larger number will be necessary, in order to keep touch and to thoroughly search the ground. These advanced scouts, however numerous, cannot be reckoned on for effective action. They must be well away, widely extended over the country, and cannot be under control. They will have to move about, often going to one side or another to avoid obstacles or examine suspicious spots. Their action, in the event of encountering the enemy, must be independent and individual. No definite instructions will apply. If suddenly met by a heavy fire in front it will generally be best to gallop back out of range. But because one scout may be thus fired on, it does not follow, as one sometimes sees, that all the others immediately come in. Let them go on to localize the enemy. It may be dangerous work, but that cannot be helped. It should be impressed on them that their work is not to fight, but to procure information. Scouts must thus work independently, and not depend too much on the action of their neighbours.

Before the battle at Modder River the cavalry scouts were mostly, if not entirely, withdrawn from the front because of some diversion on the flank, with the result that the whole force came under a sudden fire.

TRANSVAAL ARTILLERY BEFORE LADYSMITH.

OUTPOSTS AND PROTECTIVE SCREENS. 235

In rear of this fan of scouts comes a more formed body, sufficiently compact to be under command, but extended, so as not to form a target to guns unseen. *Main body of advanced guard.*

An advanced guard may have very important duties to fulfil in addition to those of mere reconnoitring, since, if an enemy does attack, it will have to hold out, not only for the time required for the main body to deploy, but often until a good position has been selected and occupied, and this may involve a pushing forward to attain suitable ground or a retreat to a better position in rear.

The distance the advanced guard should be from the main body will depend greatly on circumstances, especially on its composition and strength. If the enemy be armed with modern big guns, he might shell the head of the column from a distance of 6 miles. But this would be very dependent on the nature of the country, for though it might be comparatively rare for a road to be exposed for any length to a position 6 miles off, yet if a force were expected to be coming along a certain road, dispositions might be made by the enemy and a suitable place found so that the guns would drop a few shells among the advancing columns from such a distance. *Distance ahead of force.*

Once, while marching near Springs, we were approaching the ground selected for our camp. The advanced guard having occupied some ridges a mile or so beyond the camping ground halted, while the column gradually wound its way to the bivouac. Suddenly we were alarmed by shell after shell dropping among the transport, fired from a hill a couple of miles beyond where the advanced guard was halted.

But great care must be taken that the advanced guard of a small force is not so far from the main body as to allow an enemy, lying in wait for it, to cut in between and surround the advanced guard. At Belfast, the advanced guard of the Liverpool Regiment was cut off from the main body.

For this reason, flankers should be well thrown back on the flanks, and the main portion of the advanced guard not too far forward.

Colonel Rimington sometimes, especially when intending to effect a surprise, adopted the method of moving his column in as compact a mass as possible, with only two or three first-rate scouts and a few gallopers well out in front. These moved forward to any high ground and searched the country round with their glasses, sending back any information by the gallopers. This system was very suitable to the open rolling veldt, where ridge after ridge could be taken advantage of. The scouts when unavoidably showing themselves always rode in an opposite direction to that in which the column was moving.

Advanced guard in pursuit. With an army pursuing a retreating force, the advanced guard will have rather different duties assigned to it. If led with vigour and acumen, much may often be accomplished. The cavalry should be pushed on at all hazards, and if they can dash past or over the enemy's rear guard, the main object will have been attained, even though severe losses have to be suffered. Artillery must also be rapidly sent forward, even at the risk of coming under a close fire, so as if possible to get a chance of shelling the main body. There is comparatively little risk of an advanced guard being defeated under such circumstances, and though it may lose heavily by coming under a sudden fire, it may accomplish much if energetically led.

While the action of the advanced guard may often be to attack and endeavour to drive away an enemy in front, that of the rear guard will always be to act on the defensive. Good positions must be rapidly selected and held until the main body has retired to a suitable distance. But the positions should be chosen with an eye to a favourable line of retreat to the next place for making a stand. This, it must be remembered, need not necessarily be directly to the rear—that is, in the direction of the retreating column. *Rear guards*

For instance, if the country bordering the road be very open, and good cover, such as from the bed of a river, can be got to one side, the rear guard would best move along this. Then, if the pursuers rush along the road towards the column, they would be exposed to a hot flanking fire.

A rear guard depends greatly on mobility. It must consist mostly of mounted troops. If infantry be employed (as is usually done), and if the enemy pursue with a greatly superior force of cavalry, the chances are that the latter will be able to work round the flanks, and the slow-moving infantry will be unable to get back in time.

If the main column be moving at the rate of ordinary marching, it stands to reason that the infantry of the rear guard, if they devote any time to deployments and firing at the enemy, will gradually be left more and more behind.

A few empty mule or horse waggons will usually be available in a column, and may be used with much advantage for bringing on the infantry of the rear guard. On one occasion, during a long march on Johannesburg with Colonel Allenby's column, we were constantly attacked by parties of Boers in rear. A company of infantry was sent to assist the cavalry of the rear guard, some empty mule waggons being left to bring them on, at a trot, from one position to the next. This system worked admirably.

An attack on the rear guard can hardly be delivered as a surprise. To attack at all, while on the move, the enemy must advance at a very rapid rate to catch it up. So that anything like a serious attack would be rare, except in the pursuit of a beaten enemy.

Yet the rear guard must move extended, as guns may be rapidly brought up to get a few shells into the rear guard in hopes of harassing it, and it must be fairly strong, since (as referred

OUTPOSTS AND PROTECTIVE SCREENS. 239

to in Chap. II.) the enemy may head off the column so as to deliver a determined attack in rear.

Though, if well handled, and if in favourable country, a rear guard should be able to check the advance of, and even inflict considerable damage on, a pursuing enemy, such action is most difficult when a largely superior force of mounted troops is in close pursuit over open ground. As the pressure becomes greater, the men of the rear guard are more and more loath to dismount and fire. Their periods of retreat become longer, and their fire becomes more hurried and wild, until, as occurred in several instances during the war, the rear-guard action deteriorates into a precipitate retreat on the main body.

The duties of the rear guard under these circumstances are most important. The pursuit must be delayed at all costs, and great sacrifices must be made if necessary to save the main body from destruction. If the pursuing cavalry be enticed to come on boldly twice in succession, the third time they may be given such a lesson as will make them hesitate to advance so boldly again.

With a long column the flanks become a dangerously vulnerable point for the enemy to attack. A special feature of South African warfare was the long straggling train of ox waggons, toiling over the bad roads, necessary for the supply of a force moving far from its base. If a body be told off to push well ahead as advanced guard, and another told off to protect the rear efficiently

Flank guards.

from pursuers, a gap of many miles might occur between these two vigilant forces, and however carefully the main body might be distributed among the convoy, it was very necessary to have scouts well out on the flanks to give warning of the approach of an enemy. This is a trying duty, which also calls for mounted troops. Infantry was frequently used in this way, but it generally resulted in the flank guards gradually dropping back to the rear of the column. The route some hundreds of yards on either side of the road is almost sure to be more difficult to traverse than the road itself, especially when streams and dongas have to be crossed, and thus handicapped, infantry cannot be expected to keep up. A very few mounted men, if well away to the flank, are sufficient to give the necessary warning of attack. It is impossible to give precise orders regarding flank guards. They must use their own discretion as to reaching special objects, and to keeping their distance.

The best system for flank guards is to take up a series of positions on the flank. For instance, if a hill is seen away to one side of the road, a troop may be sent out to dismount and occupy it until the whole force has passed it; the men then mount and trot on till they get to the head of the column, and again occupy a position. Other troops meanwhile take up intermediate positions. This method allows much rest to the horses, and if attacked, one man ready in position is worth five galloping about in the open.

As soon as a force halts, advanced and rear guards become outposts, and it becomes necessary, though often difficult, to at once select good positions for occupation, in case of sudden attack.

<small>Advanced and rear guards when halted.</small>

At Vlakfontein, General Dixon's rear guard, consisting of 230 yeomanry, two guns, and a company of infantry, were halted, when the view became obstructed by clouds of smoke from a large grass fire. Suddenly a large body of Boers, estimated at 1200, dashed upon them; and the troops, not being allotted to selected positions, were defeated, and the guns temporarily captured.

CHAPTER VII.

ARMS AND ARMAMENTS.

Assimilation of the three arms. WE have hitherto been accustomed to speak of the three arms as consisting of cavalry, artillery, and infantry; mounted infantry being regarded almost as a fourth arm. But the exact line of demarkation between them is becoming less and less clearly definable. The infantryman transported by other means than his feet, whether by cart, by bicycle, or by horse, becomes practically a mounted infantryman. The cavalryman, armed with a rifle and deprived of his sword or lance (as he has lately been), and instructed to depend chiefly on musketry in action, is really the same. Even the artillerymen have lost some of their distinctiveness, having been occasionally used as cavalry for chasing the scattered guerillas, and very possibly, in the future, circumstances may occur when they can be most profitably used temporarily in this manner. The desirability of arming gunners with rifles to defend their guns and thus act as infantry was also frequently proved. Then, again, the cavalry, though generally mounted on horses, were liable to all dismounted duties, such as trench digging, outpost and garrison work, and a large proportion, through lack of horses, often marched on foot. Infantry, even though quite untrained, were frequently mounted on horses when such were obtainable, and whole battalions were, at short notice, converted into mounted infantry. The introduction of machine-guns has formed a connecting link between the rifleman and the gunner. As regards armament, uniform, and equipment, the three arms, in the later stages of the South African war, became practically identical.

It may be questioned whether such an assimilation of different branches of the service is ever likely to be introduced again. But it is undoubtedly a sign of the times, and the connection between the various classes of military forces will, in the future, certainly be closer than it was formerly.

Cavalry and Mounted Troops.

The nature of the country in South Africa, open veldt, with no hedges and ditches, and but few big rivers, formed a most favourable terrain for mounted troops. But if, in future, we have to fight in country enclosed with hedges and ditches and walls, or with thick bush or large areas of cultivation, mounted troops will probably not be so much in request.

ARMS AND ARMAMENTS.

Cavalry and M.I. The question of the relative position of cavalry and mounted infantry has been brought into great prominence during the late war. It is a much-vexed question, and one on which opinions greatly differ.

The cavalryman is supposed to be a well-trained man on a well-trained horse. A mounted infantryman, if permanently employed as such, may be as well trained, but is not supposed to be so well mounted. But horses, as we have clearly learnt, rapidly become used up, and even cavalry have to be mounted on any such material as is obtainable. Whatever little difference there may be, however, in their personnel and equipment, the tactical employment of the respective arms was hitherto supposed to be quite distinct and different, but now they seem to be merging into one.

It used to be accepted that the duties of cavalry were twofold—(1) Reconnaissance, and (2) Charging and engaging the enemy by shock action. Mounted infantry, on the other hand, were to do neither of these things, but were merely to ride fast to a given position, and then to dismount and act as infantry.

But we have now seen how in actual practice, first mounted infantry were very largely used, indeed, as often as cavalry, for reconnoitring duties, and secondly, that cavalry were very seldom employed for shock action, and that mounted infantry were occasionally used for this purpose, fixing bayonets and using them as lances.

Scouting is now both more difficult and more necessary than ever it was, so that every available mounted man will be needed to assist. And it must be agreed that, taking all circumstances into consideration, many of our mounted infantry, and this includes yeomanry and colonial mounted troops, became practically as good as the regular cavalry in this duty.

As regards the shock action of cavalry, instances in the late war of a successful charge are few and far between. It may be urged that this is no criterion of what may occur in other wars, since the Boers had no cavalry proper to operate against us, and had no infantry for us to operate against. But, on the other hand, the country was peculiarly suitable to cavalry tactics, being so open. Yet again and again, on both sides, a defeated army has retired practically unmolested. *Shock action.*

According to previously recognized rules of tactics, a body of cavalry was always to be kept in hand during an engagement, with the object of following up the beaten foe. Yet this was very seldom done during the late campaign, and the following reasons may account for the omission. With smokeless powder and long-range weapons reconnaissance is so difficult and so arduous that all mounted troops are usually fully occupied in searching out the enemy's position and trying to gain his flanks. Long ranges also imply wide distances to be traversed, and so by the end of the day the cavalry is generally fairly played out.

Then, again, the increased power of defence will enable a very few men forming a rear guard to keep pursuers at bay.

Though there may still be occasions on which a cavalry charge against demoralized infantry might have a good effect, yet they will be of rare occurrence, and it will always be very risky work, since three or four determined men with magazine rifles could do tremendous execution among a compact mass of charging cavalry, who would form such a target as could scarcely be missed. The cavalry would come under fire at such immense distances too, that infantry and guns a very long way off could practically stop a charge on their broken comrades. The country most suitable for cavalry charges, *i.e.* perfectly open, is just that which favours long-range musketry. General Rimington, as a true cavalry soldier, believes that a cavalry charge should usually be successful. "The fact is," he says, "only the *leading* charging men and horses are hit, the others meanwhile get in. Lieutenant Oliver, 6th Dragoons, charging with a troop, was killed with seventeen wounds in him, not another man hit. Rifle fire does not stop charging horsemen, and they don't see the other men fall, so they go on and don't know their own casualties till long after." Moreover, "if under a heavy shell or rifle fire at the same time, riflemen are inclined to keep their heads down and not shoot at all. I have often had to dispose some five hundred men hastily in a position of defence, and all I can say is I should not like to have been galloped at whilst a heavy fire was kept up."

On one occasion, near Ermelo, the advance of a British force was much delayed by a number of the enemy occupying some open hills to our front. It was most difficult to estimate the strength of the opponents; but the officer in command, concluding they could not be really numerous, sent forward the Inniskillings at a gallop. About thirty or forty Boers rose on their approach and tried to fly before them, but were mostly struck down or captured. In this way the road was quickly cleared, while, if we had advanced slowly and cautiously, a very considerable time would have been taken up, and our casualties probably much greater than they were. On the other hand, had the line of snipers turned out to be more numerous, or backed up by other bodies, it is quite probable that the cavalry would have suffered very severe losses and accomplished no good.

Cavalry may, too, often do most useful work by *threatening* extended infantry, and causing them to run together to form groups. If this be done by prearrangement with the artillery, the latter may have some splendid little targets to fire at.

It is certain that in future cavalry will be more and more required to act on foot as riflemen. The immense power of modern small arms demands this. Cavalry pursuing broken infantry will often come across obstacles and unfavourable ground, *Dismounted cavalry.*

and will then be still able to inflict much damage by dismounting and opening a heavy fire.

Cavalry *v.* Cavalry.
As regards the action of cavalry against hostile cavalry, it would seem that a distinct change in the methods of the past must now be adopted. Whereas formerly it was a golden rule for cavalry *never* to receive a charge at the halt, but to gallop forward to meet it, now it would almost invariably be preferable to dismount and to receive the charge with a volley of musketry. In the former case, the two forces would be on a more or less equal footing, while in the latter, a large number of the attacking force would be bound to fall before they got near their opponents.

One of the very few instances in the war of a body of horsemen attacking a similar body (in April, 1902) is thus described—

"At seven o'clock on the morning of Friday last Colonel Kekewich, moving in the neighbourhood of Rooival Junction, Hartz River, and Trekspruit, was assailed by a strong force of Boers under Commandant Kemp, numbering considerably over 1500.

"Our scouts were out a good distance in advance, as usual, with the force moving up a gentle rise.

"No sooner had the scouts topped the rise than the enemy, who were evidently waiting their arrival, and were in a long line in close formation, galloped through them shouting.

"The scouts were unable to stop the progress of the enemy, who came forward right on Kekewich's main body.

"Our men, immediately dismounting, received the attack by pouring in a fire at short range.

"So impetuous was the attack that some of the enemy reached to within forty yards of our firing line.

"The enemy turned and galloped round Kekewich's flank into the veldt, leaving on the ground 16 dead bodies and 43 wounded, and about 30 prisoners, also two guns and a pom-pom."

A charge by the Boers at Rooival had a very similar result.

It may even be possible to fire without dismounting. The Boers often did this while galloping along.

Even against artillery, times have also changed, quick-firing breech-loading guns being able to fire so rapidly as not to give the charging cavalry anything like so good a chance of "getting home" as formerly. *Cavalry v. Artillery.*

Under such circumstances it may be better for the cavalry to dismount when within close range and open fire, than to charge right up to the guns. In the latter case the gunners or escort may inflict much damage as the cavalry come up, and by forming groups among the guns may avoid being damaged themselves, and in this way the charge may absolutely fail. On the other hand, if the cavalry are in great numerical superiority, they could, by their concentrated fire, make quite sure of capturing the guns.

The true mounted infantryman is he who is *temporarily* put on a horse. If he be a well-trained *Mounted infantry.*

rider, and especially a trained horse-master, so much the better. But, in any case, he thereby becomes possessed of greater mobility. Numbers of odd horses are usually to be picked up, either having escaped, or been captured from the enemy, or belonging to inhabitants. These, though often unsuitable as cavalry remounts on account of youth or age, breed or training, may all assist in the mobility of infantry, and if a proportion of the latter are trained to do so, they can at once utilize the possession, even though it may be quite temporary. At Belmont, a number of ponies were captured during the course of the action, and a few infantrymen at once mounted on them and utilized forthwith.

Artillery.

The most important characteristic of artillery is that it can inflict damage at a much greater distance than can other arms. It also possesses some other advantages over the rifle—the burst of the shells indicates the accuracy of the shot, and faulty aim can thereby be corrected; it is able to penetrate and destroy walls, buildings, and parapets. It also possesses a certain amount of moral effect.

The greatly increased range of rifles and the invisibility of their discharge, urgently demanded a greater range in field guns, and the Boer artillery, including high-velocity, long-range Krupp and

"LONG TOM" BEFORE MAFEKING.

ARMS AND ARMAMENTS. 251

Creusot guns sighted to 9000 yards, came as rather an unpleasant surprise to us. It is certainly somewhat humiliating to be shelled by guns from a distance quite beyond the range of one's own artillery, as occurred at Ladysmith and elsewhere. Yet the many instances, on both sides, of troops having to submit to such bombardment, and yet holding out, and without serious loss, proves it not to be of great importance. But it must also be remembered that South Africa is, as a rule, a country peculiarly suited to long-range artillery fire, on account of the remarkably clear atmosphere, the absence of trees, the frequency of high ground commanding wide views. So that here, again, we must not be led away with the idea that long-ranging guns are an important feature of modern war. In England, for instance, it is only rarely, owing to weather and to the profusion of large trees, that troops could be seen manœuvring more than two or three miles off.

As regards the moral effect of artillery fire, it has certainly proved to be of less importance than was to be reasonably expected. Though instances could be quoted where, after a musketry duel had been in progress for some time, a shell or two has at once decided the contest in favour of those possessing the artillery, yet, on the other hand, the effects of a heavy bombardment on the nerves of the defenders has often proved most disappointing. Experiences in the several sieges show how callous one can become of bursting shells. Even

Moral effect.

the terrific explosions of lyddite often failed to dislodge the Boers from their rocky fastnesses.

Material effect. It has been pretty clearly proved that the actual damage done by artillery fire, especially against troops under fairly good cover, is not great. Percussion shell, as usually used by the Boers, has but very little effect, and even a well-burst shrapnel may only be equal to a volley from a couple of hundred rifles (but without the penetrative power).

At Driefontein, two Boer Creusot-Schneider guns, bursting shrapnel accurately at 4000 yards, enfiladed our infantry lines, but the losses sustained were quite insignificant.

Perhaps one of the most surprising lessons gained in the war is the harmlessness in the results of artillery fire.

The affair of the armoured train at Chieveley according to Mr. Winston Churchill's account seems almost incredible to the theorist. He says: "The Boers had opened fire on us at 600 yards with two large field guns, a Maxim firing small shells in a stream, and from riflemen lying on the ridge." Yet the train passed on safely till derailed by an obstruction. Then "the Boer guns, swiftly changing their position, re-opened from a distance of 1300 yards . . . and a third field gun came into action from some high ground on the opposite side of the line. . . . Besides the three field guns, which proved to be 15-pounders, the shell-firing Maxim continued its work," and after bombarding the train for "seventy minutes by the clock," the engine was able to steam away to safety!

Even if we may have to make a large allowance for the exaggerations of an ardent correspondent the results seem surprising. The Boer gunners were considered good marksmen, and during this time one would suppose at least 300 or 400 15-pounder shells would have been discharged. Would any artillerist before the war have ventured the statement that these four guns, at as close a range as 1300 yards, firing for over an hour, could not damage a stationary locomotive sufficient to disable it, or, that of the 120 men on the train, not more than about one-third were wounded?

This, by the way, looks promising for the utility of armoured motor-cars in war.

So that, on the whole, artillery fire, except under special circumstances, such as the bombardment of a town or building, would be of little use were it not that it can act at distances far greater than musketry.

It must, then, be recognized that the importance of artillery fire has not increased to the extent that some sanguine gunners seemed to expect, and that the effects of lyddite, of automatic and quick-firing guns, and of very large field pieces has not handicapped the relative progress of rifle fire.

It has been clearly shown that the artillery duel is now by no means a necessary introduction to a battle. The Boers generally kept some, at all events, of their guns "dark" until the attack was

Artillery preparation.

far developed. It is usually impossible to ascertain the exact position of the guns of the defence, especially if they are drawn back until required. It would certainly usually be the case, if the defenders were weak in artillery, that they would not attempt a duel with a more powerful enemy, and disclose the position of the guns, but would reserve their fire until a good target presented itself.

Artillery has been proved not to be so efficacious for preparing the way for infantry attack as was hitherto supposed. But it is still useful for keeping down the hostile fire *during* the attack. A matter requiring great care is to know, at a distance of perhaps 3000 or 4000 yards, when to cease firing owing to the attacking troops having gained the position. There were many instances of infantry storming a position only to find themselves under a heavy artillery bombardment from their own side. At Stormberg, a number of the Irish Rifles fell victims to our own gun-fire. At Magersfontein, the remnants of the Highland Brigade might have succeeded in taking the trenches but for the fire of our guns pouring on them.

The numerous instances of our gunners firing on our own troops imperatively calls for some remedy, and large and powerful telescopes (which could easily be carried on the guns) would seem the only means of obviating such fatal occurrences.

Howitzers are especially useful for preparing for infantry attack, as, owing to the high angle at which they fire, infantry can advance in safety while the projectiles are discharged over their heads, and there is not much fear of premature bursts.

The searching of the far slope of a hill with shrapnel, especially at the moment of general retreat, was on many occasions most effectual. **Indirect fire.**

Often, too, it was found desirable to shell places quite invisible. Pienaars River station, situated in the midst of dense bush, was held by the Boers. We approached with caution, being sniped among the trees. Soon, from a clearing, a windmill was visible over the tree-tops two or three miles away. Our guides reported that the Boer laager was a few hundred yards to the north of the windmill. Our guns opened fire in this direction, and the force advanced to find that the enemy had precipitately retired on several shells falling close to camp.

It has often been laid down as a maxim that artillery cannot come into action within 1000 or 1500 yards of hostile infantry. This is generally very true when referring to a large battle, but it is all dependent on the exact circumstances. If the infantry is not numerous, and the gunners behind some cover, the latter need not suffer so greatly as to be obliged to withdraw. Also, if the riflemen are distracted by an attack or demonstration from another quarter, they will probably be too much engaged in that direction to devote their attention to the distant guns. Artillery was often brought up within 600 or 800 yards of the enemy with great effect. **Proximity of hostile small arms.**

Such a situation as that of Colonel Long's guns at Colenso is very different. Here were, apparently, several thousand rifles concentrating their fire on the guns in the open.

Locating guns. The introduction of smokeless powder renders it *exceedingly* difficult to locate an enemy's guns. The Boers, who used both kinds of powder, when firing with smokeless are said to have posted a man behind cover away to one side of a gun with some charges of black powder, so that when the latter was fired he lit the powder, and the cloud of smoke attracted our fire to the wrong spot.

For locating the enemy's artillery there is again need of powerful glasses.

Yet another lesson we have learnt from our late enemies is the desirability of rapidly changing the position of guns. They often fired three or four rounds, and then, when our guns began to concentrate their fire on that spot, moved off rapidly, and took up a new position a hundred yards or so to one flank.

One often speaks of guns being "silenced," implying that they have been rendered useless and no longer capable of doing damage. But it was a frequent practice among Boer gunners, when the fire became hot, to temporarily abandon a gun and retire under cover. After a due interval they could usually return and again open fire. Had our gunners more often followed these tactics their losses would undoubtedly have been less, and the temporary cessation of artillery fire, though not

Fort Wylie. Lieut. Roberts's Grave. Hlangwani.

COLENSO: SPOT WHERE THE GUNS WERE LOST.

desirable, may be very preferable to heavy losses. If the guns at Colenso had at once been temporarily abandoned, they would probably not have been permanently lost.

A quick-firing gun discharging nine rounds a minute is as good as four guns firing only two, and it is better in many ways than the four, needing fewer men, fewer horses, and less transport for their food and forage.

Quick-firing guns.

Yet a single gun forms a better mark for concentrated fire, and once damaged its fire is stopped and not only slackened. But guns are not often actually struck.

This war is the first in which "automatic" artillery has been used, and "pom-poms," first introduced by the Boers, rapidly became a favourite arm. They possess several most valuable characteristics. The little shells indicate very clearly the range, and if seen to fall short or otherwise off the mark, the aim can be instantly altered and more shells immediately fired. This is invaluable against a moving object, and these guns will form one of the greatest enemies to cavalry in bodies. They would also be most useful against intrenchments in knocking over the head-cover, and therefore good for preparing for attack. They also, if able to reach the necessary angle, should prove fatal against balloons.

Pom-poms.

The rapid discharge of shell after shell has a considerable moral effect. Even the succession of reports is terrifying to them in front. It has often

R

been said that its "bark is worse than its bite," and the actual effect of pom-pom fire is small. This may be so, but I happen to have seen considerable execution done with it. At Modder River all the men of the machine-gun detachment of my battalion were killed or wounded, later one of our officers was wounded by it, and shortly after a man killed. In a later fight we saw a Boer hit in both legs by a shell.

Escort to guns. It has hitherto been considered a necessity for guns to have a special infantry escort told off to defend them in case of need. But often a gun is suddenly ordered to take up a position some distance off, and, not to waste valuable time, it is trotted off, leaving its tired infantry far behind. Probably, soon after the escort has got up, orders will come for the gun to return and continue the march with the main body, and the escort will again be left to struggle after the column. Colonel Long at Colenso trotted his guns far away from their protecting infantry, with fatal consequences. Artillery escorts should, if possible, be formed of mounted troops. Machine-guns, perhaps mounted on the limbers, might assist.

The position to be taken up by an escort has to be carefully considered. In Colonel Benson's unfortunate fight at Brackenlaagte two guns were placed on rising ground, "where there was a hollow and sharp dip to the west of the two guns, coming to within 20 yards of the cannon. For some inexplicable cause the three companies of the Buffs were halted a considerable way in rear of the guns upon a reverse slope," with disastrous results.

Infantry.

The foot soldier has the great advantage of being very independent. He is comparatively inconspicuous, and can at once make himself still more so by crouching or lying down; he presents a smaller vulnerable target than a horseman; he can easier surmount fences, walls, and other obstacles; and he does not require vast quantities of forage to be taken in his transport, needing but one-tenth the amount of food and water that a horse does.

His great disadvantage as compared with the horse soldier, is his slowness in getting over the ground.

Since the strength of a force is really dependent on the amount of fire it is capable of delivering, a few men armed with very rapid-firing weapons are as good as a large number with ordinary arms. It may be said that with rifles capable of expending a large amount of ammunition in a short time, a great difficulty will arise in the supply. But this really only refers to the actual delivery to the firing-line. As regards transport, the food supply for one man is infinitely greater in weight than his ammunition supply. Thus, if an arm be adopted firing twice as quickly as the old one, only half the number of men could deliver the same amount of

Quick-firing rifles.

fire, and one-half the weight of rations formerly carried would enable the new force, with the same transport as before, to carry a far larger supply of ammunition per rifle.

Infantry formations. Since infantry can so rapidly and easily make themselves less conspicuous and present a smaller vulnerable target by lying down, a golden rule to impress on men is, "when in doubt, *lie down*." It has often been the case that in sudden emergency men start running about and crowding together, causing confusion and a good target.

On active service one should *always* be prepared in case of attack. For this reason troops should always be extended as much as possible. The old-fashioned system of crowding men together in quarter-column and suchlike formations is most objectionable. They are very convenient in a barrack square, but are quite out of place in the field. Such masses form good targets not only for aimed fire but for a chance shell; they prevent the men using their weapons at once, and if a panic occurs, such as I have unfortunately witnessed, the men become entangled in a hopeless crowd. Column or line is good enough for all purposes.

Against cavalry. With the great increase in rapidity of fire, infantry would now depend more upon fire action than on their bayonets for warding off cavalry, and if they will only *lie still* when threatened, and shoot at the advancing horsemen, they need have no fear of the result. It may be they even have to fire above the heads of their comrades around, but so long as all are lying down, no harm need come of that. It would be waste of precious time to run in to form groups, and the latter, moreover, would form targets for artillery and especially pom-pom fire.

Carts, especially in countries intersected with roads, should form valuable means of transporting infantry. This method of conveyance has two great advantages over the riding-horse: (1) The men need not have any special training or knowledge to be so employed; (2) a fewer number of horses can convey a given number of men (a horse is supposed to be able to draw four times as great a weight as it can carry on its back). About 15 men can be comfortably taken in a 10-mule waggon. But this method is practically useless for scouting work—for which mounted troops are so urgently required—and gives a good target if they come under fire. *Infantry in carts.*

The question has often been discussed as to whether small calibre machine-guns are to be classed as artillery or musketry. Maxim guns on wheeled carriages drawn by horses were attached to infantry battalions during the war. They were then supposed to act as a group of infantrymen. But they formed an object every bit as conspicuous as a field-gun and yet accompanied the firing-line! This was soon recognized as an absurdity, and the machine-guns relegated to the rear. The one gun of the battalion was also found to draw a concentrated fire upon itself. A Maxim on a tripod mounting *Machine-guns.*

is hardly more visible than an ordinary rifleman, and if such guns be sufficiently numerous they will not draw special fire upon themselves.

Machine-guns are specially useful for the protection of a convoy. While marching with a small convoy near Hornies Nek, we were attacked by a number of Boers concealed in the grounds of a farmhouse. Our weak escort of mounted troops seemed likely to be driven in, but it so happened that we had with us a Maxim gun which had been sent into Pretoria for repair. This was rapidly got out and brought into action with most telling effect, and our assailants soon cleared.

Fire-tactics.

Fire-arms all-important. As no improvements have lately been made or are likely to be made in hand-to-hand weapons, while great progress has been introduced in fire-arms, the latter have now become of supreme importance, and are likely to continue to progress in the future, while other factors, such as natural qualities, training, means of progression, etc., will probably not change to any great extent. Therefore tactics are now entirely subordinate to power of fire, and the strength of a force is to be measured by its capability of discharging projectiles.

Fire-tactics, then, are the tactics of the battle-field, and have been already gone into in considering "Attack" and "Defence." Here has been pointed out the importance of a steady hail of fire discharged in the direction of an enemy, and that this is even more effective than aimed fire. At close ranges the latter may have good effect, but when firing at distances over 1000 yards, under the circumstances of war, the chances of hitting a small target aimed at must be very remote. A marksman taking plenty of time over each carefully aimed shot is not more likely to hit his mark than a man rapidly discharging a dozen shots in the general direction.

The best general policy will therefore be to keep up a hail of fire on the position where the enemy is likely to be, or even on where he is likely to come to.

Targets may be said to be of two kinds, *visible* **Target.** and *material.* An instance of the first would be a conspicuous fortification, easily seen from afar, but which is not likely to suffer much if it is hit. The second may be a battalion in compact formation under cover of some trees, quite invisible at a distance, but every bullet that came that way might find a billet. The tendency in modern war being to render men, guns, and even fortifications inconspicuous, implies that this invisible, material target will be that which we shall most often want to get at. It may seem strange to the old-fashioned warrior to think of two large forces, like those facing one another at Modder River, lying on the ground and firing away for hour after hour without any visible and vulnerable target whatever.

This uncertainty renders the choice of a target a difficult matter. Often under such circumstances a single head appearing will be the signal for an outburst of fire from all sides. Any visible object will attract the fire. But this desire to fire at *something* is not always to be encouraged. A hat on a stick has often caused a waste of ammunition. A couple of horsemen galloping away in the distance are often fired at, not only by riflemen, but even by artillerymen, with very little chance of any damage being done.

There may, of course, be circumstances under which such firing may be advantageous, such as when it is desirable to keep off prying scouts, when a shell dropped near them, or a long-range volley, may cause them to retire.

All that can be done is to carefully consider and deduce from known facts whereabouts the enemy is, and then to shower bullets on that locality.

It is most difficult, under present circumstances, to find a good target for artillery. In attacking a position, if the defenders reserve their artillery fire, no target is presented, and all that the attacking guns can do is to promiscuously sprinkle shells about the position on the chance of some taking effect.

Expenditure of ammunition. As a rule, then, rapidity of fire is all-important. The greater the mass of bullets discharged in the direction of the enemy the more chance there is of some doing damage. The future development of the rifle will probably be in the direction of automatic loading weapons, like the Mauser pistol.

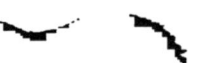

The supply of ammunition will necessarily be a great difficulty, but it is one that must be faced, as parsimony in this respect will undoubtedly be fatal to efficiency.

In several actions men were known to fire over 200 rounds a-piece.

But though long-range fire may be typical of modern fighting, it must not be forgotten that combat at close quarters is still not only possible, but is often of supreme importance. On Waggon Hill some of the troops "were within 100 yards of the enemy for 15 hours exposed to a deadly fire" (Off. Desp.). At Spion Kop, and many other places too, close-range fire settled the day. *Short-range fire.*

Volleys undoubtedly have a great moral effect. As concentrated fire on one spot has a greater effect than a disseminated fire along the whole line, so while men may get accustomed to occasional bullets striking here and there, a sudden volley of bullets pattering around may just give that shock to their nerves that may cause them to decide not to remain to experience another such shower. So that if the enemy appears to be faltering a volley may just turn the scale. *Musketry volleys.*

Since the concealment of a position from the enemy is now of the greatest importance, every effort should be made not only to hide from view guns and men, but also to puzzle the enemy as to their exact position. By firing over interposing rising ground or from some distance behind trees, an enemy may be deceived as to where the fire *Concealment.*

comes from. The reports both from guns and rifles may then be deceptive, and the reports of firearms are now practically the only guide to their whereabouts. Possibly in future we may have noiseless firearms, an innovation which would have almost as great an effect as that of smokeless discharge.

It is these developments of the future that we must look out for, and our tactical training must be in accordance with our experiences of the most modern introductions, instead of, as so frequently practised in the past, taking as our basis occurrences of thirty or more years ago.

APPENDIX

A TYPICAL POSITION.

Exemplifying Difficulties of Attack and Uncertainties of Defence.

Attacking Force (p. 47)—

    Intelligence reports enemy in position on hills Y Z.

    Scouts sent forward report coming under fire, at A, from farmhouse X.

    Other scouts fired on from R (*vide* p. 94).

    No further information obtainable. (Trenches invisible and position indeterminable.)

    Guns moved up and thoroughly shell house and grounds at X (though really unoccupied, defenders being intrenched away to sides, *vide* p. 186).

    Infantry advance to attack. On crossing ridge AB come under long-range fire, and are henceforth practically beyond control of commander.

    Right flank hangs back on coming under desultory fire from W.

    On getting to CC, line comes under heavy enfilade fire from W, also from guns at V.

    It is only now, when all the defenders open fire, that a true idea of the position is realized. But no orders can now be conveyed to the advanced line, which will be suffering heavily.

Defending Force (pp. 76, &c.)—

    Expecting attack along road.

    Reserves at U supposed to be safe, come under shell fire from B.

    Defenders' artillery unable to reply.

    Advanced post at R, after firing a few rounds and puzzling enemy as to disposition of defenders, retires along river bank.

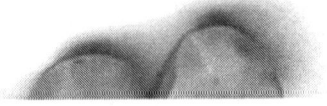

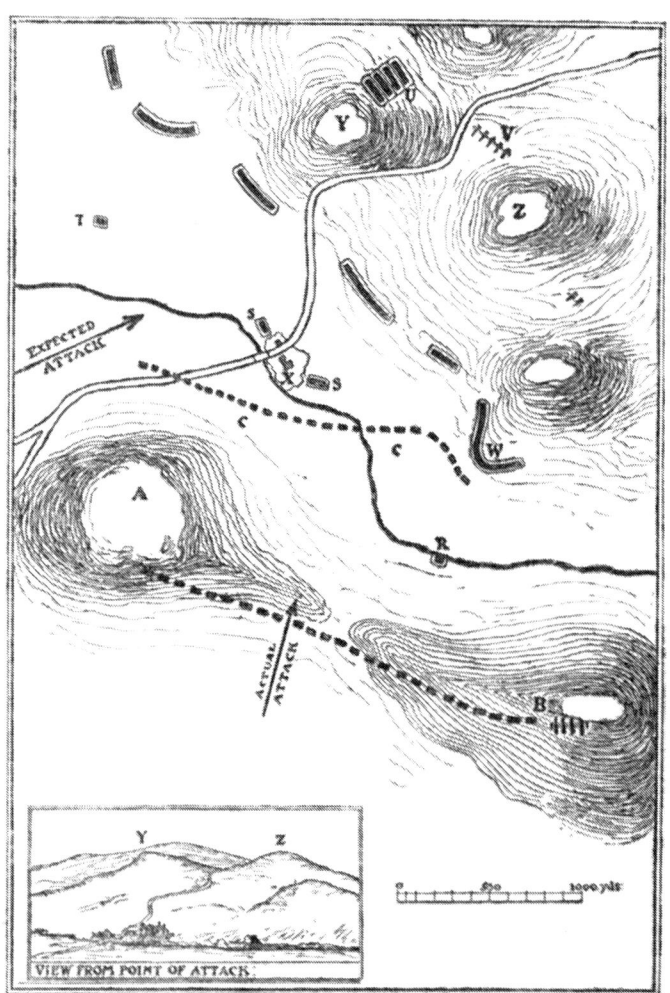

A TYPICAL POSITION.

INDEX

INDEX.

A.

Advance, independent of base, 26–29; rapidity of most important, 36; method of, 56–57; short rushes of alternate units, 56–57; tactical instances of, 70; danger of hurried movements of, 70

Allenby's, Colonel, column, 238

Ambuscades, important results of, 94–95

Armoured train, guns firing at, 252

Armoured waggons, use of, in defence, 186; construction of, 188

Arms and armaments, assimilation of the three arms, 242–243

Army Corps, arrival of Sir Redvers Buller's, in South Africa, 17; his division of, 19

Artillery, preparation unnecessary of, 50; artillery and musketry preparation, 50-51; long-range fire of, 51; power of, 71; field of fire for, 131; characteristics and advantages of, 250; Krupp and Creusot guns, 250–251; moral effect of, 252; material effect of, 252–253; harmless results of artillery fire, 252–253; artillery preparation, 253–254; indirect fire, 255; proximity to hostile small arms, 255–256; locating guns and silenced guns, 256;

quick-firing guns, 257; pom-poms, valuable characteristics of, 257; escort to guns, 258

Attack and defence, relative value of, 42, 76

Attack, when likely to succeed: conditions stated, 45–46; objective in, importance and usual plan of, 47; most favourable hour for, 49; front line of, 56; final stage of, 62; different methods of, 64; attack in woods, 65; by mounted troops, 71; fortification in, 72; details of, 73; counter, 101–104; best form of, 103

Attacking a force on line of march, 66; three conditions of, 67; example of from Boers, 68

Attacking a camp, best time for, 69

Attacks, instances of successful, 46

B.

Baden-Powell, General, his force hemmed in at Rustenburg, 38; 175

Barbed wire, uses of, in war, 193–197

Basuto, a, at Colesberg, 225

Bayonets, 62; Boers fear of, 63; charge of, when advantageous, 102

INDEX.

Beatson, General, 229
Belfast, 112, 191; advanced guard at, 236
Bell-tents, objections to, 191; a mistake for outposts, 228
Belmont, battle at, 21, 46; shelling of, 50; mistake at, 61; outposts at, 203, 250
Berkshire Regt., the, 49
Blockhouses, near Brandfort, 79; usual dimensions and make, 178; objections to, 178; improved, 179; covered-in trenches best form of, 180
Bloemfontein, occupation of, 23; Lord Roberts's advance from Modder River to, 26; Cronje's march towards, 35; works at, 195
Boers, at Ladysmith, 17; at Magersfontein, 21; defective strategy of, 22–23; and Lord Roberts's advance from Modder River, 24; tactics of, in Natal, 32
Brackenlaagte, 68; Colonel Benson at, 258
Buffelsfontein, Imperial Light Horse at, 74
Buffer-post, 214
Buller, Sir Redvers, arrival in South Africa, 17; divides his forces, 19; reference to, in Lord Roberts's despatch, 27

C.

Camp, attacking a, 68–69
Campbell, Major, 101
Cannon Kopje, 147
Cape Colony, 26
Carleton, Colonel, his force captured, 34
Cavalry and Mounted Troops, 243–245; duties of cavalry, 244; of mounted infantry, 244; use in scouting, 245; shock action of, 245–247; dismounted cavalry, 247–249; against hostile cavalry, 248; Colonel Kekewich's fight with Commandant Kemp, 248; mounted infantry, 249–250
Chrissie Lake, 93
Churchill, Mr. Winston, 252
Colenso, 44, 96; trench at, 157; gunners at, 182; Colonel Long's guns at, 256, 257, 258
Colesberg, 18, 19, 81, 225
Columns, small mobile, success of, 28; easily marred, 28; independent columns desirable, 31
Commander, good view for, 130
Communication, lines of, 29; importance of, 29; use of railways for, 29
Communication, means of, 188
Communication trenches, 188
Concentration and extension, 30; South African experience of, 31
Concentration of garrison undesirable, 82–85
Country, hilly, 123; importance of knowledge of, 127
Cover, important considerations of, 77–78; when attacking commanding ground, 115; artificial cover important, 142
Cradock, 226
Cronje, headed off Paardeberg, 33; retires from Magersfontein, 35

D.

Daily Telegraph, account of affair near Bethel, 229–231
Defence, power of increased, 42, 43; two factors of, 77; position for, importance of clear field of fire, and cover, 77; small forces in, 78–79; disposition of force for, 80; misleading enemy, 81; concentration of garrison undesirable, 82; allotment of troops in, 85; semi-permanent position in, 89; distribution of garrison in, 90; long lines of, 91–92;

INDEX.   275

advanced posts in, 94; ambuscades, 94; concentration of fire in, 95; reservation of fire in, 96; necessity of look-out, 97; supports and reserves in, 98; standing to arms in, 100; counter attack in, 101; best form of counter attack, 103; evil of retirements, 104; permanent positions, 127-132; temporary positions, 132, 133; defence of towns and villages, 183-186; three lines of, 183; at Mafeking and Rustenberg, 184

De Wetsdorp, retirement from, 25; disaster of garrison at, 68, 133

Driefontein, attack on, 46; Creusot-Schneider guns at, 252

Dummy-works, 190

Dundee, 24; retreat from, 34, 38

E.

Edenburg, trenches at, 176

Elands River, 24; attempted relief of, 49, 80

Enemy's guns, undesirable positions for, 131

Ermelo, 247

Evolution of strategy and tactics, 15

Excavations and erections, 172-173; at Mafeking, 173

Extension of a force, 30; reasons for, 53, 54

F.

Fauresmith, 80

Field intrenchments, two classes of, 150

Fire, "firing line" a misnomer, 56; when to open, 57, 58; concentration of, 58-95; diagram of, 83; amount of, 88; reservation of, 96; field of, illuminating, 190

Firearms, improvements in, and added power of defence, 19, 22, 23

Fires, on outpost, 225

Fire-tactics, 262-266; firearms all important, 262; policy regarding, 263; two kinds of targets, 263; importance of rapidity of fire, 264; ammunition, 264, 265; concealment, 94, 265, 266; future developments of, 266

Firman, Colonel, 204

Flank protection, importance of, 58

Fog, characteristics of, 40

Food supply, necessity of concentration for, 30

Force, advantages of a small, 31, 32; attacking on the move of a, 66, 67; conditions of such attack, 67; surrounding a, 69

Formation of force for attack, 55

Fort Prospect, 80, 141

Fortification in attack, 72, 73, 141-145; old and new systems of, 142, 145; invisibility of works, 145-147; important adjuncts to, 187

G.

Game-Tree Fort, assault on, 175

Garrisons, distribution of, 90; use of civilians in, 185

Gatacre, 18

Glencoe, 24

Gloucester regiment at Nicholson's Nek, 104

Grasspan, 21, 46, 121

Ground, advantages of knowledge of, 49; selection of, for attack and defence, 109; shape of hills and undulations, 109-113; advantages of occupying commanding ground, 113-114; detailed consideration of advantages, 114-118; supposed advantages of high hill positions investigated, 121-122; advantage to defence of

INDEX.

open country, 124; tactical importance of woods, 126
Guards, advanced, rear, and flanking, 231-233; general principles of action, 232; chief duties, 233-234; main body of advanced guard, 235; advanced guard in pursuit, 236-237; rear guards, 237-239; attack on, 238; duties of, 239; flank guards, 239; best system, 240; when halted, 241
Gun-pits, 180, 183; at Magersfontein, 181; usual pattern of, 182; machine-gun-pits, 183
Guns, 250, etc.; position in camp of, 100

H.

Hart's Brigade, 158
Hartz River, 248
Head-cover, importance of, 162-164
Hedges and walls, for defence, 186
Heidelburg, trenches at, 176
Helvetia, 79, 229
Hills, shapes of important, 109, 113
Holnek, 136; trenches at, 176
Hornies Nek, 262
Horses, holding of, 72; good training of Boer, 72; grazing for, 132
Housing of men, 191; objections to bell-tent, 192
Howitzers, 254
Hurried movements, undesirable, 70

I.

Illuminating field of fire, 190
Illustrations, concentration of fire, 83; profiles of hills, 110; position at base of slope, 120; diagonal shelter-pits, 152; position of a trench, 157; narrow loopholes limit fire, 160; semi-circular all-round trench, 162; head-cover, 164; notched parapet, 167; plan of loopholes, 168; covered-in trench, 177; a typical position, 269
Infantry, advance of, 50
Infantry attack formation, 52; alteration of, older drill-books discarded, 52; invisibility of, 54; front line of, 55-56
Infantry, 259-262; independence of, 259; quick-firing rifles of, 259-260; formation of, 260; against cavalry, 260-261; in carts, 261; advantages of this method of conveyance, 261; machine-guns as attached to, 261-262
Inniskillings, the, 247
Intrenching, importance of, 142
Invaders and defenders, opponents in war classified as, 16; advantage of assailant, 16-17; his advance from one point or several, 18
Invincibility of small garrisons: uselessness of investing such, 24
Irish Rifles, the, 25
Itala Fort, 80, 141

J.

Jamestown, 79
Johannesburg, 238

K.

Kaalfontein, 80, 110
Kekewich, Colonel, 248
Kemp, Commandant, 248
Kimberley, 18, 19, 24, 25, 80
King's Royal Rifle Corps, 2nd Batt., 102
Klerksdorp, attack on Colonel Von Donnop's convoy near, 68, 79
Klipdrift, attack on Lord Methuen's force at, 68

INDEX. 277

Koster's River, bushmen at, 115
Kuruman, 79, 80

L.

Ladybrand, 24, 80, 131
"Lady Roberts" gun, 229
Ladysmith, 17, 21, 24, 38; blowing up of Long Tom at, 39; sickness at, 89; attack on, 101, 142, 251
Lee-Metford bullet, 164
Leeuwberg, 111
Lichtenberg, siege of, 24, 80
Lincolnshire Regiment, the, 25
Lines of defence, advantage of long, 21-23; *e.g.* the Boers at Magersfontein, 21
Liverpool Regiment, the, 236
Long-range weapons, their effect on tactics, 58
Look-out posts, necessity of, 190
Loopholes as affecting trace, 159-160; importance of splay of, 159; arrangement of, 161; size of, 164-165; best shape of, 166; construction of, 167-171; three methods of splaying, 168-169; sand-bag loopholes, 170; number of necessary, 171

M.

Machadodorp, 210
Machine-guns, 50, 261
Mafeking, siege of, 24, 25; General Baden-Powell's irregular corps at, 38; attack on Boer trenches at, 39, 65, 80, 85, 87; advance to, 70; Eloff at, 81; defence of, 97; successful counter attacks at, 103; water supply at, 128-130; trenches at, 187; hospital at, 188
Magersfontein, the Boers long line of defence at, 21; Cronje's retirement from, 35; checks on our night march to, 39; an example of defence, 44; shelling of, 50, 63, 72; trenches at, 74; Cronje at, 90; Boers at, 91, 96; the Highlanders at, 97, 103; water supply at, 128; description of fine position at, 139, 141, 217; Highland Brigade at, 254
Majuba Hill, capture of, 65
Maritzani, 70, 85
Maxim guns, 252, 262
Men, housing of, 191
Methuen, Lord, 18, 21, 27
Middleberg, 136
Modderfontein, 79
Modder River, position at, 21; Lord Roberts's advance from, 24, 26; Lord Methuen at, 27; Cronje on the, 35; advance on, 52, 55; orders at, 64; situation at, 70; making cover at, 72; strength of Boers at, 88; reservation of fire at, 96, 97; position at, 117, 120; standing camp at, 206; scouting at, 234; pom-poms at, 258; opposing forces at, 263
Moedwil, 192
Molopo River, 70, 85
Monument Hill, 155
Mooi River, 187
Mules, their use for transport, 28
Muller, Commandant, 229, 230
Musketry, greater effect than artillery, 51; good field of fire for, 129, 262
Mutual assistance, its necessity for concentrated attack, 18; its difficulty with divided forces, 19; difficulty lessened by long range of modern firearms, 20

N.

Natal, 17, 18, 19, 21, 22, 24; Boer tactics in, 32; our slow movements in, 36, 233

INDEX.

Nicholson's Nek, capture of Colonel Carleton's force at, 34; troops short of ammunition at, 46; official despatches at, 84, 103

Night marches, reasons for, 38; difficulty of, for large bodies of troops, 39; advantage of small bodies for, 39

Nooitgedacht, General Clements at, 69, 217

Northern Bush-veld, obstacle at, 195

Northumberland Fusiliers, the, 25

O.

Observation posts, 131, 190

Obstacles and obstructions, overcoming of, 73; construction of, 192; to advance of enemy, 131; importance of artificial, 192, 193; barbed wire as, 193; a fence or abattis as, 193; how to put up barbed wire for, 193; hidden posts, 194; piles of stones as, 194; wires drawn across trees as, 195; desirable qualities of a wire entanglement as, 196; wire netting as, 197; position of, 197; blocked roads as, 198; destroyed bridges as, 198; ditches as, 199; ruined railways as, 198; impassable fords as, 199

Obstructions, their use in delaying advance or retreat, 198-199

Oliver, Lieutenant, of 6th Dragoons, 246

Outposts and protective screens, losses through neglect of, 200; outpost duty, 201; objects of outposts, 202-204; strength of, 204; extent of front, 205-208; disposition of troops on, 208; suitable positions for, 214, 215; position by night of, 215-218; standing to arms of, 225; best method of posting of, 218; extract from *Daily Telegraph* on, 229

P.

Paardeberg, Cronje headed off at, 33, 36; assault on, 44; Canadians and Gordons at, 73; repulse at, 88, 129, 141

"Parados," 145

Patrols, abolition of, 221; duties of, 221-222; objections to, 222-223; substitute for, 224

Picquets, groups of, 210; distance apart of, 211, 212

Pienaars River, 39; move to, 66; capture of British outpost at, 98; abattis round, 197, 255

Piet Retief, difficulties of General French's column at, 28

Plateau, advantages of position on a, 134

Poplar Grove, infantry at, 52, 55; Boer trenches at, 187

Position, captured, what to be done with, 75; semi-permanent, 89-90; selection of, 113-123, 127-132; selection of temporary, 132-133; occupation of, 133-135; risky to occupy unknown, after dark, 134; key of, 133-135; varies with object, 135; principal objects of, 135-140; for long occupation, 135-137; for temporary, 137-140

Posts, ability of small, 78; examples for and against, arrangements of, 92; protected from each other's fire, 93; advanced, uses of, 94

Pretoria, how Boers might have opposed our advance on, 22, 38; electric bells and electric light at, 190; wire entanglement at, 195, 262

Prinsloo, surrender of, 34

Pursuit, a good opportunity of, 35

INDEX. 279

R.

Railways, most efficient means of supply, but vulnerable, 29
Reddersburg, troops short of ammunition at, 46
Retirements, a mistake: untenable positions should be relinquished in time, 34; troops disheartened by, 35; give opportunity of pursuit to enemy, 35
Roberts, Lord, his arrival in Africa, 18; his successful advance, 20; his tactics, 24; his advance from Modder River to Bloemfontein, 26; his despatch, 27; his delays succeeded by rapid movement, 36
Roosenekal, night march on, known beforehand, 37
Rustenberg, evacuation of, 35; General Baden-Powell's force hemmed in at, 38

S.

Sand River Bridge, 80
Sanna's Post, 117, 122
Scouts, uses of, 56
Screens, of bushes, etc., use of, 188; bullet-proof, 188
Sentries, duties of, 218; tendency to sleep, 219; objections to challenging, 219; inlying, 227-231
Shells, shrapnel, 50; descent of bullets, 164; harmless, 252
Shock tactics, not essential, 33
Small and large forces, advantages of former, 31-32
Small parties, work done by, 50; methods of attack, 50
Soudan campaigns, square formation of, 70
South Africa, necessity of concentration for supply in, 30; difficulties of campaign in, 31; cause of reverses in, 37

South African Constabulary, 176
South African War, new ideas derived from, 16
Spion Kop, 44, 75, 119, 134, 147, 265
Springfontein, Major Baden-Powell at, 213
Springs, 235
Standing to arms, objections to, 225
Stormberg, 18, 37, 44; unfortunate affair at, 49, 64; Irish Rifles at, 254
Strategical advance and tactical defence, 32-33, 69
Strategy, how affected by recent changes, 40-41
Support, to isolated forces, necessity of, 25-26
Supports, 58; position of, 59, 60; defence of, 98
Supports and reserves, 226-227
Surface, details of, 123; summary of characteristics of, 124-126
Surprises in war, effect of, 37
Surrounding a force, 69

T.

Tactics, advantages of defensive, 76, 77
Talana Hill, instance of successful attack on, 46, 98; description of, 112; massing of troops at, 126
Tod, Lieutenant, death of, 102
Trace, or plan of works, important consideration, 158; affected by loopholes, 159
Trekspruit, 248
Trenches, 147, 148; tools available for making, 148; details of, 149-152; long *versus* short, 153-154; position of, 156-157; best, semicircular, 161-162
Troops, allotment of, deserving more attention, 38; mounted, attack by, 71, 72; allotment of, 85; considerations of, 86; in

INDEX.

camp, 99–100; standing to arms of, 100
Tugela, the, 22; Sir Redvers Buller on, 27, 53
Tweefontein, 191, 219

U.

Uitval Nek (*see also* Zillikat's), surrender of Greys and Lincolns at, 25, 100

V.

Vaal, the, what Boers might have done at, 22–23
Victorians, 229
Viljoen, Ben, 229–230
Vlakfontein, General Dixon's rearguard at, 241

W.

Waggons, armoured, use of, in defence, 186

Wagon Hill, 101, 142, 265
Warmbaths, abattis round, 197
Warming, means of, 189
Water supply, 128, 137
Weapons, long-range, 58
Wepener, 24, 80, 87, 142
White, Sir George, hemmed in at Ladysmith, 17, 24
Willow Grange, 121
Winburg, 80
Woods, defence of, 126
Works, independence of, 155, 156

Y.

Yule's, General, retreat from Dundee, 232

Z.

Zillikat's Nek (*see also* Uitval), troops badly posted at, 46; re-occupation of, 49, 198
Zuurfontein, 80; trenches at, 176; means of communication at, 188

THE END.

LaVergne, TN USA
31 March 2011
222313LV00002B/144/P